Corrosion Management and Control In Desalination:

AN INDUSTRIAL GUIDE

Roger Francis and Abdelkader Meroufel

AMPP™

AMPP
Association for Materials Protection and Performance

Printed in the United States of America

Print ISBN: 978-1-57590-422-1
eBook ISBN: 978-1-57590-390-3

AMPP
Association for Materials Protection and Performance
15835 Park Ten Place
Houston, TX 77084
ampp.org

Dedication

To my wife, for her endless patience while I retired to the office to write yet another book.

— RF

This effort is dedicated to my parents for their sacrifices, my wife for her support, and to humble Professor Anees Malik who introduced me to corrosion in desalination with his 27 years of accumulated experience.

— AM

Contents

Preface

We have been active in all types of desalination plants for many years, either solving corrosion problems, or, less often, preventing them from happening. There are very few publications that discuss the specific corrosion problems in desalination plants, and there are none written with the plant engineer in mind. Plant engineers often do not have an in-depth knowledge of corrosion, and because of this, they need help with day-to-day problems that may involve corrosion.

Modern desalination plants are often designed with a 40-year lifespan, so it is important that plant engineers know how to tackle corrosion problems to maintain water production without shutdowns or, in extreme cases, shortening plant life. We conceived the idea for this book as being a helpful guide that can sit on an engineer's desk, and which will give advice not just on diagnosing corrosion failures, but on how to monitor the plant's resistance to corrosion by inspection and monitoring. This book has been compiled from our experiences, but we would like to thank Glen Byrne, Steve Clarke, Charles Desportes, CAPCIS Intertek, Ineos Composites, and Saline Water Conversion Corporation for additional technical information and for the use of some of their images. Thanks are also due to Rolled Alloys, Intertek, and Steve Clarke for permission to use some of their photographs. We would also like to thank Ian Francis who did the original line drawings and designed the cover.

We hope you find this book useful.

—Roger Francis and Abdelkader Meroufel
March 2022

1 Introduction

The world's population is steadily increasing and with it is an increasing demand for freshwater, for both drinking and irrigation. In many areas of the world, particularly in arid regions, there are limited freshwater sources such as rivers and groundwater, so desalination is being increasingly used to produce freshwater to satisfy both requirements. Although desalination is sometimes carried out on brackish waters and highly saline well waters, most desalination plants generate fresh water from seawater with a daily production that reached 95 million m^3/day in 2019.[1]

There are three main processes used in desalination plants, the oldest of which is multistage flash (MSF), where the water is essentially boiled at low pressure and the steam that flashes off is condensed and is post-treated to meet drinking water quality specifications. The second process is multiple effect desalination (MED), in which low pressure steam is used to force evaporation of seawater and the vapor is then condensed for drinking water. Although largely used on land, small capacity MSF and MED units have been fitted to large ships, such as cruise liners, to generate fresh water.

The third process is seawater reverse osmosis (SWRO), where chloride salts are selectively removed from seawater by forcing it at high pressure through a semipermeable membrane. This method involves no heat transfer but requires enough electricity to power the high pressure pumps that are required. All three of these methods have advantages and disadvantages, as discussed for example in Reference 2.

MSF plants require a large quantity of steam and are usually associated with a power station, which can generate extra steam and electricity for the desalination plant. They can produce very large quantities of fresh water and have been widely used in the Middle East, where they are associated with fossil fuel (oil or natural gas) fired power stations. They consume relatively large amounts of energy to generate fresh water and may not be competitive with MED or SWRO in some applications.

MED plants were traditionally smaller than MSF plants, but only require a supply of low pressure steam. With improvements in scale control chemicals, large plants of a similar size to MSF have become possible with a lower water production cost.

SWRO plants require no heat, but they do require chemical pretreatment and a significant quantity of electricity to drive all the high and low pressure pumps. Some recent plants have constructed large solar arrays in the desert or have installed wind farms so they are powered by renewable energy. They do have the advantage that they can be constructed in a wide range of sizes and can be modularized, so extra trains (i.e., desalination lines) can be added as water demand increases. They are also popular at holiday destinations where water demand varies through the year, because SWRO is relatively easy to turn on and off. Most of the new desalination plants currently being designed and built are SWRO.

All three processes can be subject to corrosion unless suitable mitigation practices are implemented, such as a cost-effective materials selection for the many components, and they are operated within their normal limits of use. Some of these corrosion types are common to two or all three types of desalination, while others are specific to one method. In this book, detail is not given about corrosion mechanisms per se, but it provides references where this information can be found. The main focus is on the common major corrosion problems seen in desalination plants and methods of avoiding them. The book also gives information on measures to monitor corrosion and avoid corrosion when the plant is nonoperational.

In Chapter 2, the various metal alloys used in desalination plants are discussed, including their properties and any special problems, particularly those associated with welding and fabrication. The main properties of the common polymers used in desalination plants are presented, along with the importance of selecting the correct grade of polymer with different aggressive fluids. Chapter 3 addresses galvanic and atmospheric corrosion—the two common corrosion types found in all types of desalination plants—and ways of avoiding them.

In Chapter 4, the common corrosion problems seen in the various parts of MSF desalination plants and the most suitable materials of construction are discussed. Chapter 5 covers the same thing for MED desalination plants. Chapter 6 discusses the common corrosion problems seen in the three main parts of SWRO desalination plants: the low pressure section, the high pressure section, and the brine discharge, as well as the most common materials for construction. The effects of possible future developments in membrane technology on materials of construction are also discussed. Chapters 4, 5, and 6 include case histories demonstrating some common corrosion problems.

All desalination plants spend some time under nonoperating conditions, either during commissioning or shut-down for maintenance. Chapter 7 discusses the possible corrosion problems that can occur under such conditions and methods of avoiding corrosion and future failures.

Chapter 8 discusses corrosion mitigation practices through design, materials selection, and by controlling the environment. It addresses material selection for corrosion resistance, thermal conductivity, strength, and weldability.

In Chapter 9, the typical lifetime of the main components in all three types of desalination plants are discussed. The life of most materials is variable, depending on the local conditions and how well the plant is maintained. In the series of tables given, material lifetimes can be used to select options for both new build to last the life of the plant, or for short-term repair, where the plant's remaining life is shorter.

Chapter 10 discusses the main ways of assessing corrosion in desalination plants, by both direct and indirect assessment. The problems with corrosion probes are discussed, particularly for detecting localized corrosion.

Chapter 11 addresses standards, quality, and procurement, and highlights the need for additional requirements over and above those of ASTM International(1) standards if material of adequate quality is to be procured. Appropriate testing of the different alloys is discussed. Some recent failures due to poor quality material are discussed, highlighting this need.

The alloys are referred to by their generic names, but the compositions to Unified Numbering System (UNS) designations[3] are used throughout and ASTM standards are used to discuss the various product forms. Test methods are described to either ASTM or International Standards Organization (ISO) standards, as appropriate. The equivalent alloy grades to other standards, such as European Standards (ENs), can be found on several comparative websites on the internet. The Appendix contains the nominal composition and generic names of all the cast and wrought alloys discussed in this book. This is so that an alloy composition can be readily checked when reading any of the chapters.

If this book is read from beginning to end, it will be noticed that there is a certain amount of repetition, including definitions for all acronyms. This is because it is expected that most readers will dip into sections relevant to their current problem and they will not wish to keep following references to other chapters. Where the point at issue requires more lengthy discussion, however, references to other chapters have been made.

1. ASTM International was formerly known as the American Society for Testing and Materials.

References

1. Jones, E., M. Qadir, M.T.H. van Vliet, V. Smakhtin, S-M. Kang, "The State of Desalination and Brine Production: A global outlook," *Science of the Total Environment* 657 (2019): pp. 1343–1356.
2. Francis, R., *Avoiding Corrosion in Desalination Plants* (Houston, TX: NACE International, 2018).
3. ASTM, "Metals and Alloys in the Unified Numbering System" (Warrendale, PA: SAE International; and West Conshohocken, PA: ASTM International, 2004).

Bibliography

ASTM International, West Conshohocken, PA, USA

ISO (International Standards Organization), Geneva, Switzerland

2 Materials

This chapter discusses the materials commonly used in desalination plants, and where to find the appropriate properties when designing a component or part of a plant. The main focus is on the welding of metals and fabrication of polymers because these are areas where problems can arise if appropriate precautions are not taken, particularly when fabricating onsite.

2.1 Parent Metals

In this document a range of metals and alloys are discussed with reference to their use in desalination plants. The nominal compositions, Unified Numbering System (UNS) numbers, and the generic names of these alloys are shown in the Appendix, and reference should be made to this to identify any alloy discussed in the text.

Where details of an alloy's mechanical properties are required, it is recommended that standards such as those from ASTM International be consulted, e.g., ASTM A240 for stainless steel sheet and plate. It is important to consult the standard for the appropriate product form because the properties can vary slightly with product form for some alloys.

For other properties, such as thermal conductivity or thermal expansion coefficient, these can usually be found in manufacturers' data sheets. One final factor to check is availability of a particular material in the desired product form because not all alloys are available in all product forms, e.g., super ferritic stainless steel, which is only available as thin sheet and heat exchanger tube.

A brief description of the properties of the alloys in the Appendix can be found in Reference 1.

2.2 Welding

The present section aims to provide a brief overview on the critical welding aspects that induce typical corrosion problems that may occur in desalination plants. This includes welding consumables, heat practices, and necessary quality control (QC) tests. When joining metals by welding it is important to select the correct filler metal to avoid galvanic corrosion of the weld metal (see Chapter 3, "Galvanic Corrosion"). In addition to this, the weld Procedure Qualification Record (PQR) will need additional testing for some alloys, over and above those from the American Society of Mechanical Engineers (ASME), or whatever fabrication code is being used. This section summarizes the suitable fillers for some of the common alloys used in desalination plants, as well as suggesting additional tests to ensure the PQR will produce corrosion resistant welds. Table 2.1 shows some common filler metals for the parent metals discussed below. Table 2.2 shows the American Welding Society (AWS) designations for different filler wire types.

TABLE 2.1 Common filler metals for gas shielded welding of the alloys (discussed in Sections 2.2.1–2.2.4).

PARENT METAL		FILLER METAL		COMMENT
Generic Name	UNS Number	Generic Name	AWS Designation	
Carbon Steel	-	High Ni Filler	E7018-G-H4	High nickel to avoid weld corrosion
ACI[a]	F43000	Ni/Fe	EniFe-1	Difficult to weld; PWHT[b] essential
90/10 Cu-Ni	C70600	70/30 Cu-Ni	ERCuNi	Needed to avoid iron coming out of solution
70/30 Cu-Ni	C71500	70/30 Cu-Ni	ERCuNi	Matching filler
		Alloy 400	ERNiCu-7	Sometimes used
Al-Bronze	C61400	Al-Bronze	ERCuMnNiAl	Iron content prevents cracking
NAB[c]	C63200	NAB	ERCuNiAl	Matching filler; PWHT essential
		Al-Bronze	ERCuMnNiAl	No PWHT, but may lack strength
304L	S30403	304L	ER304L	Matching composition
		309	ER309L	Fully austenitic weld
316L	S31603	316L	ER316L	Matching composition
		309	ER309L	Fully austenitic weld
317L	S31703	317L	ER317L	Matching composition
904L	N08904	Alloy 625	ERNiCrMo-3	Needed to maintain corrosion resistance
6%Mo	S31254 N08367 N08926	Alloy 625	ERNiCrMo-3	Needed to maintain corrosion resistance

TABLE 2.1 (cont.) Common filler metals for gas shielded welding of the alloys (discussed in Sections 2.2.1–2.2.4).

<table>
<tr><th colspan="2">PARENT METAL</th><th colspan="2">FILLER METAL</th><th rowspan="2">COMMENT</th></tr>
<tr><th>Generic Name</th><th>UNS Number</th><th>Generic Name</th><th>AWS Designation</th></tr>
<tr><td rowspan="2">Lean Duplex</td><td>S32101
S32202
S82011</td><td>2109</td><td>ER2101</td><td>Normal filler with extra nickel</td></tr>
<tr><td>S2304
S32101
S32202
S82011</td><td>2209</td><td>ER2209</td><td>Higher alloy for higher corrosion resistance</td></tr>
<tr><td>2205</td><td>S32205</td><td>2209</td><td>ER2209</td><td>Extra nickel to maintain phase balance of weld</td></tr>
<tr><td rowspan="2">Superduplex</td><td>S32760</td><td>Z100X</td><td>ER2594</td><td rowspan="2">Extra nickel to maintain phase balance of weld</td></tr>
<tr><td>S32750</td><td>2509</td><td>ER2509</td></tr>
</table>

a. ACI = Austenitic Cast Iron
b. PWHT = Postweld heat treatment
c. NAB = Nickel aluminum bronze.

TABLE 2.2 Weld filler wires and their AWS designations.

<table>
<tr><th rowspan="2">ALLOY</th><th colspan="2">FILLER WIRE</th><th rowspan="2">TYPE</th><th rowspan="2">COATING/ FORM</th></tr>
<tr><th>Designation</th><th>AWS Number</th></tr>
<tr><td rowspan="2">NiFe</td><td>ENiFe-1</td><td>A5.14</td><td>TIG</td><td>Wire</td></tr>
<tr><td>ENiFe-C1</td><td>A5.15</td><td>MMA</td><td>Basic flux</td></tr>
<tr><td rowspan="3">904L</td><td>E385-16</td><td>A5.14</td><td>MMA</td><td>Rutile flux</td></tr>
<tr><td>E385-15</td><td>A5.14</td><td>MMA</td><td>Basic flux</td></tr>
<tr><td>ER385</td><td>A5.9</td><td>TIG</td><td>Solid</td></tr>
<tr><td rowspan="3">309</td><td>E309L-17</td><td>A5.4</td><td>MMA</td><td>Basic flux</td></tr>
<tr><td>E309-16</td><td>A5.4</td><td>MMA</td><td>Basic flux</td></tr>
<tr><td>ER309L</td><td>A5.9</td><td>TIG</td><td>Wire</td></tr>
<tr><td rowspan="2">Lean duplex</td><td>E2101-15</td><td>A5.4</td><td>MMA</td><td>Basic flux</td></tr>
<tr><td>E2101-16</td><td>A5.4</td><td>MMA</td><td>Rutile flux</td></tr>
<tr><td rowspan="4">22%Cr</td><td>E2209-16</td><td>A5.4</td><td>MMA</td><td>Rutile flux</td></tr>
<tr><td>E2209-17</td><td>A5.4</td><td>MMA</td><td>Rutile flux</td></tr>
<tr><td>E2209-15</td><td>A5.4</td><td>MMA</td><td>Basic flux</td></tr>
<tr><td>ER2209</td><td>A5.9</td><td>TIG</td><td>Wire</td></tr>
<tr><td rowspan="2">Superduplex</td><td>E2594-15</td><td>A5.4</td><td>MMA</td><td>Basic flux</td></tr>
<tr><td>ER2594</td><td>A5.4</td><td>TIG</td><td>Wire</td></tr>
<tr><td rowspan="2">625</td><td>ENiCrMo-3</td><td>A5.11</td><td>MMA</td><td>Basic flux</td></tr>
<tr><td>ERNiCrMo-3</td><td>A5.14</td><td>TIG</td><td>Wire</td></tr>
</table>

2.2.1 Carbon Steel

With carbon steel weldments, the most frequent corrosion problems may occur in the form of preferential corrosion of the weld metal or the heat-affected zone (HAZ). Usually, the former results from an improper choice of welding consumables. High-silicon, low-nickel consumables yield welds that rapidly corrode in seawater, while the reverse (i.e., low-silicon, high-nickel) can produce welds that are acceptable. Figure 2.1 shows severe corrosion of a carbon steel weld in low-oxygen (~100 ppb) seawater, which was 50 mV more-negative in potential compared with the parent pipe. A common filler used to avoid preferential weld corrosion contains 0.6–1.0% nickel (Ni) and 0.4% copper (Cu).[2]

FIGURE 2.1 **Preferential weld corrosion of carbon steel in piping carrying partially deaerated seawater (~100 ppb).** ***Photo courtesy of Intertek***

Preferential corrosion of the HAZ has been observed when welding is carried out with a heat input that is too low. Then, the HAZ shows higher hardness and a susceptibility to initiate preferential corrosion in some aqueous media at room temperature. The preferential HAZ corrosion in seawater has been attributed to the presence of low temperature transformation products (martensite, lower bainite, or retained austenite).[3] The solution in this case is to weld with a higher heat input to produce a softer HAZ. However, this can be a problem with some low-alloyed steels where the welding produces a coarser grain size than occurs in the parent metal, which reduces the strength compared to parent metal.[3]

For more information, Bond presented a paper reviewing the corrosion of carbon and low-alloy steel welds, their causes, and the mitigation methods.[2]

2.2.2 Austenitic Cast Iron (ACI)

As discussed in detail in Chapter 4 on MSF distillation, austenitic cast iron is susceptible to stress corrosion cracking (SCC) in warm seawater or brine. Then, the presence of stresses is totally undesirable. Site welding of this alloy is likely to generate detrimental residual stresses. It is unlikely that staff doing site weld repairs would have the necessary experience to do it properly. Even with an experienced fabricator in a welding shop, however, it is not easy and the results can be variable.

Because of the SCC risk, it is recommended that the component be stress relieved after welding. This is usually done with a postwelding heat treatment (PWHT) at around 450–550°C, followed by slow cooling. If ACI fails in service, it is often simpler to replace it, or upgrade to a more corrosion resistant alloy.

2.2.3 Copper Alloys

For some of the copper alloys, the selection of a filler metal is very important. Copper and its alloys can be welded by all the common arc welding techniques. It is important that the weld metal is not anodic to the parent metal because the weld bead has a much smaller area than the parent metal, so it is necessary that its potential is either the same as the parent metal or more positive, particularly for the parts that will be in contact with ambient temperature aerated seawater. If this is not achieved, rapid, localized attack of the weld metal can occur.[4]

The copper-nickel alloys are often used in seawater and welding is preferred for high-strength joints. The iron content of 90/10 copper-nickel (Cu-Ni) exceeds the solubility limit at room temperature, so iron may come out of solution if a matching composition filler is used and the weld cools too slowly. This can reduce the corrosion resistance of the weld metal, particularly the erosion-corrosion resistance.[5] For this reason, it is common to weld 90/10 Cu-Ni with 70/30 Cu-Ni filler. The higher nickel content (typically above 29%) greatly increases the iron solubility limit and there is no risk of iron precipitation during cooling. Thin sections of 90/10 copper-nickel have been welded with matching filler because the weld is small and cools sufficiently quickly to keep the iron in solution.

The 70/30 Cu-Ni (UNS C71500) alloy is usually welded with a matching composition filler. This produces welds that are not anodic to the parent metal so they will not corrode preferentially. Alloy C71500 has been welded with alloy 400 filler, but this is generally unnecessary because it is more expensive than matching filler and does not generally confer any significant corrosion resistance benefits.

The welding of the aluminum (Al) bronzes is somewhat more complex, particularly for the multiphase alloys. The subject is discussed in detail by Meigh[6] and by Spiller.[7] Here, a brief overview is presented and readers wishing more information should consult these references.

One problem with welding the aluminum bronzes is the ductility dip. Meigh[6] gives recommendations for minimizing the risk of cracking, involving both the choice of welding parameters and the selection of the filler composition.

When welding the single phase, less-ductile aluminum bronzes (alloys containing ≤ 9% Al and 2–3% Fe), it is usual to use matching composition filler. The alloy containing 8% aluminum and 3% iron is sometimes preferred because iron increases the strength of the weld metal and reduces the risk of cracking development during the cooling stage.[6]

When welding nickel aluminum bronze (NAB), it is common to use a matching filler. However, the weld bead cools very quickly and not all the beta phase present at high temperature may transform to alpha and kappa phases. Beta phase is highly anodic to alpha phase and will corrode rapidly in seawater. Hence, a postweld heat treatment is desirable to ensure that all the beta phase transforms. The most common heat treatment is that specified by the Royal Navy (United Kingdom) for all welds to NAB. This involves heating at 675°C for six hours followed by air cooling. Where PWHT is not possible, then the use of Cu/8Al/3Fe filler is an alternative. Although no postweld heat treatment will be required, the joint will not be as strong as one made with matching filler and may not be as strong as the parent metal. The use of a single phase filler does not result in any preferential weld corrosion in seawater because the potentials of both aluminum bronzes are similar. Figure 2.2 shows a seam-welded NAB pipe that had not been postweld heat treated. After exposure to seawater, the beta phase corroded rapidly with the result that the seam weld "unzipped."

FIGURE 2.2 Corrosion of the beta phase in an NAB seam welded pipe that had not been postweld heat treated.

Provided that they have been postweld heat treated, NAB welds do not generally cause a corrosion problem in seawater.

2.2.4 Stainless Steels

The most commonly used stainless steels in desalination plants are austenitic and duplex stainless steels.

2.2.4.1 Austenitic Stainless Steels

The low-alloy austenitic stainless steels, such as 304L and 316L, are usually welded with matching composition filler metal by any of the arc welding processes. Austenitic stainless steel welds typically contain 2–20% of delta ferrite. While this phase reduces the risk of hot cracking during the weld solidification, it may induce a dissolution of the passive film in deaerated acidic solution as reported by Pujar et al.[8] If there is a concern about delta ferrite in the weld metal, then a higher nickel filler can be used to ensure a fully austenitic weld metal. Such an alloy type is 309, for which UNS W30913 (nominally Fe/23Cr/12Ni) and W30920 (nominally Fe23Cr/12Ni/2Mo) are the fillers for 304L and 316L, respectively. The 309 filler is often used to join 300-series stainless steels to carbon steel to avoid excessive delta-ferrite formation due to dilution of the weld metal.

One advantage of delta ferrite is that it can make the weld metal more resistant to SCC, because of the duplex microstructure. The minimum delta ferrite content to achieve this has not been determined, but a value ≥ 10% would seem reasonable.[9]

As mentioned above, solidification cracking can affect austenitic stainless steel welds. This issue is discussed in detail by Gooch, and he concludes that filler metals that produce 5–10% delta ferrite largely eliminate the risk of solidification cracking in well-made welds.[10] Even with fully austenitic weld metals, the risk of cracking can be drastically reduced by using a low travel speed during welding.[10]

Where it is desired to ensure that a 300-series stainless steel weld is not susceptible to intergranular corrosion, such as occurs when carbides precipitate, it is recommended that an intergranular corrosion test to ASTM A262 method E be conducted.[11] It is advisable to carry out the test in the as-welded condition.

Higher alloy austenitic alloys, such as 904L and the 6%Mo alloys, suffer from dendritic segregation when welded with matching composition filler. The dendrites in the weld metal are depleted in chromium and molybdenum, greatly reducing the corrosion resistance. Hence, it has become standard practice to weld these alloys with a nickel-based consumable, such as alloy 625, where the chromium and molybdenum content in the dendrites is still acceptable for most applications in desalination plants.

One feature of welding with fillers, such as alloy 625, is the creation of an unmixed zone at the weld root. This is where the parent metal has melted and then refrozen without mixing with the filler metal. This area has been shown to have a much lower corrosion resistance.[12] By careful choice of joint design and welding parameters to ensure better mixing of the weld pool, it is possible to minimize the unmixed zone, although it is difficult to prevent its formation completely.[12] The use of an even higher alloy filler has no effect on unmixed zone formation.

When stainless steels are being used beyond their limits of use, it is common to weld overlay with a more corrosion resistant alloy, and alloy 625 has been widely used for this in seawater. When the component is creviced, such as a flange face, then it has been demonstrated that the crevice corrosion resistance of as-welded 625 is not as good as that of parent 6%Mo alloy, because of the dendritic as-welded structure.[13] To get an increase in the crevice corrosion resistance, it was necessary to use a higher alloyed nickel-base filler such as alloy 59, 686, or C-22.[13]

When qualifying a weld procedure for 904L or 6%Mo austenitic stainless steel, it is advisable to carry out a corrosion test on the welded joint. The usual test is an ASTM G48 ferric chloride test at 40°C for 24 hours. A "pass" is no pitting when viewed at X20 magnification and a weight loss of $< 4\ g/m^2$. This ensures that the weld will have acceptable resistance to corrosion in seawater.

2.2.4.2 Duplex Stainless Steels

Autogenous welding of duplex stainless steels is not generally recommended because the weld metal will be highly ferritic because of the fast cooling rate, and there will be a significant reduction in corrosion resistance and toughness.

If duplex alloys are welded with a matching composition filler, then the weld will be ferritic, as with autogenous welding. Only a full postweld solution anneal heat treatment will completely restore the microstructure and give good properties.

However, most welding of duplex stainless steels requires their being used in the as-welded condition. The filler wires for this requirement are usually matching in composition with an extra 2–4% nickel. Common examples are 2209, 2509, and Z100X. The extra nickel enables sufficient austenite to form during the fast cooling of the weld so that the 50/50 austenite/ferrite phase balance is maintained. With the superduplex alloys, ASME rules now permit welding of 2507 and Z100 with either of the two alloys' fillers.

However the weld is made, and whatever filler is used, it is most important that the final weld has a good phase balance, in the range of 35–65% ferrite for preference. This will ensure both good mechanical properties and good corrosion resistance.

The manufacturers of duplex stainless steel parent metals and welding fillers will usually have publications advising on suitable joint design, gas purging, heat inputs, and interpass temperatures. It is important that any PQR for duplex stainless steel keeps to these guidelines. Going outside these can result in the precipitation of third phases (such as sigma) in the HAZ or the weld, which reduce both toughness and corrosion resistance.[14]

It is often stated that welds are not as corrosion resistant as parent metal, but if correctly made, duplex welds are often similar in performance.[15]

It is well known that if superduplex stainless steel is incorrectly welded, then sigma and/or chi phases may precipitate in the HAZ; the chromium and molybdenum depleted zone around the particles is then susceptible to corrosion. Figure 2.3 shows water leakage from the corroded HAZ of a superduplex weld in the high pressure section of an SWRO plant, due to precipitation of sigma phase.

FIGURE 2.3 Water leaking due to corrosion in HAZ of superduplex weld in HP section of an SWRO plant. *Photo courtesy of Rolled Alloys*

There are several ways to improve the corrosion resistance of duplex stainless steel welds. One is to pickle the weld with a commercial pickling paste or gel, especially formulated for duplex stainless steels.[14] This is difficult to do for small-diameter pipe welds. Another method is to use nitrogen-containing welding gases for both the shielding and backing gases.[14] Argon with 2–3% nitrogen has been used successfully in both the welding shop and in the field.

As with 6%Mo austenitic stainless steels, duplex stainless steel welds should have additional testing over and above the requirements of the fabrication code. This was first recognized by the oil and gas industry, and they drove the creation of ISO 17781, which gives the testing and pass/fail requirements for a range of duplex stainless steels.[16] This includes both parent metal and welds. These tests are just as important for the desalination industry.

The tests for lean duplex, standard duplex, and superduplex welds are shown in Table 2.3. The impact toughness test is to demonstrate acceptable toughness of the welded joint. The corrosion test shows that there are insufficient third phases to cause significant corrosion of the welded joint. Nitrides may not cause pitting, but can cause a high weight loss, and their presence can result in failure after a few months. The microsection is to determine the phase balance and whether any third phases are present. If the weld has failed either one or both of the other two tests, the microsection can suggest the cause and possible remedial action.

TABLE 2.3 Test requirements for duplex stainless steel welds (taken from ISO 17781).

ALLOY		TEST		
		Toughness	Corrosion	Microsection
Lean Duplex	Test	Charpy impact test at 20°C	ASTM A1084 for 24 h+	Polished and double etched*
	Pass	45J Min average; 35J single minimum	No pitting at X20; Wt. loss <4 g/m^2	Ferrite 30-70%; No significant third phases
2205	Test	Charpy impact test at −46°C	ASTM G48 at 22°C for 24 h	Polished and double etched*
	Pass	50J Min average; 40J single minimum	No pitting at X20; Wt. loss <4 g/m^2	Ferrite 30-70%; No significant third phases
Superduplex	Test	Charpy impact test at −46°C	ASTM G48 at 35°C for 24 h	Polished and double etched*
	Pass	50J Min average; 40J single minimum	No pitting at X20; Wt. loss <4 g/m^2	Ferrite 30-70%; No significant third phases

* Double etching = electrolytic etching first in oxalic acid and then in NaOH.

+ No temperature specified

If there is a problem passing the corrosion test in ISO 17781, Francis[14] suggests several methods of improving the corrosion resistance. The choice will depend on the components being welded (plates, pipes etc.).

2.3 Polymers

These are used extensively in desalination applications for both chemical storage vessels and piping to handle both corrosive chemicals and seawater. Their combination of chemical resistance to a wide range of chemicals, light weight, and relatively low cost, compared to Ni-Cr-Mo alloys and high-alloy stainless steels, make them a very attractive option.

2.3.1 Fiberglass Reinforced Plastics (FRPs)

FRPs are made from thermoset polymer resins reinforced with glass fibers. The glass fibers provide the strength, while the thermoset resin provides a corrosion resistant binder. While the different resins that are used for FRP have different mechanical properties, the strength of the laminate is governed by the glass content, orientation of the fibers, their adhesion with the polymeric matrix, and method of fabrication.

The structural layer of the FRP laminate comprises mainly glass to provide a high strength composite material. The surface of the laminate exposed to the process is polymer rich and is known as the "corrosion liner" or "corrosion barrier." Only sufficient reinforcement from either glass or polyester is required to strengthen the corrosion liner (Figure 2.4).

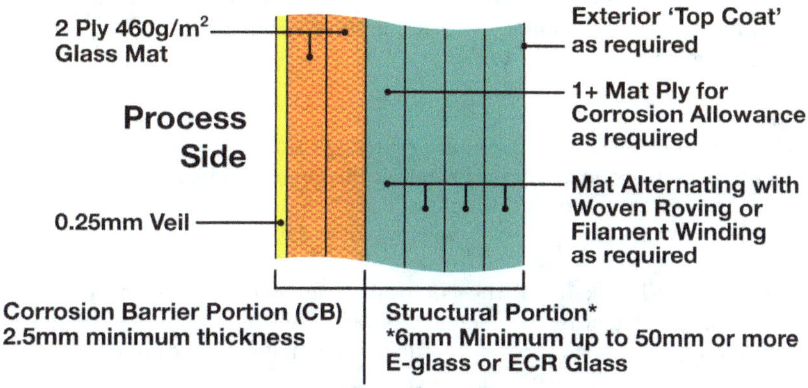

FIGURE 2.4 Schematic diagram showing makeup of FRP.
Diagram courtesy of Ineos Composites

Epoxy resins are the least expensive of the polymers used to make FRP composites. Their low cost makes them very attractive for numerous applications. They are widely used to handle seawater, particularly as large-diameter feed pipes.

Other resins are sometimes used where the liquid is particularly corrosive, the temperatures are higher, or better mechanical properties are required:

- Epoxy vinyl ester resins (higher chemical resistance)
- Epoxy novolac-based vinyl ester resins (higher temperature limit)
- Brominated epoxy vinyl ester resins (good fire resistance and chemical resistance)
- Brominated epoxy novolac vinyl ester resin (better than brominated epoxy vinyl ester resin)
- Elastomer-modified bisphenol-A (BPA) epoxy vinyl ester resin

The latter resin is typically used when bonding other FRP resins to steel or other substrates, such as concrete or wood. Their semi-elastomeric behavior, when used as an interlayer, accommodates the different thermal expansion coefficients of the FRP and the substrate.

Equipment and piping made from FRP have proven to be extremely well suited to many corrosive environments. Vinyl ester resins are immune to attack by most salts, brines, and weak acidic solutions. FRP piping can be used in caustic applications, but care must be taken to prevent caustic attack of glass fiber reinforcement. For this reason, FRP pipe for caustic applications is specified with a "synthetic" veil, which is typically polyester reinforced with alkali resistant fibers.

Chemical resistance for FRP is best assessed on a case-by-case basis. Resin suppliers publish chemical resistance guides for their products, and these can be extremely useful for selecting FRP resins for commonly used chemical environments.[17] The makeup of the corrosion liner will determine the chemical resistance and therefore the long-term performance of the laminate. The variables that dictate the makeup of the corrosion liner follow:

- Resin type
- Composition of the reinforcement (glass type)
- Composition of the outermost sheet in the corrosion liner (veil)
- Thickness of the corrosion liner
- Type of curing system used to polymerize the resin

The thickness and makeup of the corrosion liner is best assessed by an experienced person familiar with the chemical system and the types of resin that can be used. In many cases, testing of representative laminates with candidate corrosion liners can provide useful materials selection data.

2.3.2 Polyvinyl Chloride (PVC)

PVC and CPVC (post-chlorinated polyvinyl chloride) are variants of the same polymer backbone. PVC is made by the substitution of one of the hydrogen atoms on the ethylene monomer repeat unit with a chlorine atom. The resulting polymer is more chemically resistant, stiffer, and stronger than polyethylene. The mechanical properties of PVC can be changed significantly by the addition of plasticizers. Plasticized PVC can be made to be very flexible and has a wide range of applications in the packaging, clothing, and other industries. The grades of PVC used in the chemical and desalination industry are "un-plasticized" or "uPVC." This material is widely used for handling chemicals–including hydrochloric acid–across its concentration range (up to 37%). PVC can be post-chlorinated by taking the PVC resin and exposing it to chlorine under heat and pressure to produce post-chlorinated PVC or "CPVC." CPVC has improved mechanical strength over PVC and is often used where the temperature for PVC is too high. Typically, the limiting temperature for PVC is 60°C, whereas the limiting temperature for CPVC is

93 °C. Reference 18 is a useful comparison of the limiting temperatures at different pressures for thermoplastic piping materials. Note that in terms only of mechanical properties, CPVC outperforms most thermoplastics and even outperforms certain fluoroplastics, such as polyvinylidene fluoride (PVDF), over most of the temperature range.

Another issue is the brittleness of the thermoplastics. If a worker drops a wrench or other heavy object on a PVC pipe, it can shatter and empty potentially corrosive fluids over the site. For this reason, it is common to apply a thin FRP "wrap" around the thermoplastic pipe as armoring against mechanical damage. It is advisable to carry out a drop test or barcole hardness test on thermoplastics to ensure they are not excessively brittle.

2.3.3 Ultraviolet (UV) Degradation

Most polymers will suffer degradation if exposed to strong sunlight with a high UV index, as occurs in tropical climates and the Middle East. The degradation manifests itself in the form of a chalky appearance or color change at the polymer surface. The effects of this can be minimized by taking some suitable precautions.

It is important to specify polymer grades with UV protection built in through the addition of stabilizers (e.g., hindered amine light stabilizer) or pigments (carbon black, titania). However, in very hot climates this is not sufficient to give extended life, and a topcoat of white latex paint acts as a sunblock and also tends to reflect solar radiation. Many end users in hot and/or tropical climates tend to replace their PVC and CPVC pipe on a timed life basis not only because of chemical attack, but also because of UV issues.

Many chemical suppliers supply their chemicals in totes, which are high density polyethylene (HDPE) in a steel cage, and these will also need protection from UV light if the chemical is dispensed directly from such containers.

Where polymers have failed by UV degradation, it is often because a grade without UV protection has been used outdoors, or a supplier has provided a grade without the UV protection.

References

1. Francis, R., "Avoiding Corrosion in Desalination Plants" (Houston, TX: NACE International, 2019).
2. Bond, S., "Corrosion of Welded Components in Marine Environments, Prevention and Management of Marine Corrosion" (Cambridge, United Kingdom: TWI Ltd, 2003).
3. Robinson, J.L., "Preferential Corrosion of Welds," *The Welding Institute Research Bulletin* 20 (1979).

4. Francis, R., *Galvanic Corrosion: A Practical Guide for Engineers,* 2nd edition (Houston, TX: NACE International, 2017).
5. Francis, R., *The Corrosion of Copper and Its Alloys; A Practical Guide for Engineers* (Houston, TX: NACE International, 2010).
6. Meigh, H.J., "Cast and Wrought Aluminium Bronzes–Properties, Processes and Structures" (London, United Kingdom: Institute of Materials Minerals and Mining, 2000).
7. Spiller, K.R., "Practical Aspects of Welding Aluminium Bronze," *Welding and Metal Fabrication* (November 1971): p. 401.
8. Pujar, M.G., R.K. Dayal, T.P.S. Gill, and S.N. Malhotra, "Role of Delta-Ferrite in the Dissolution of Passive Films on the Austenitic Stainless-Steel Weld Metals," *Journal of Materials Science Letters* 1, 8 (1999): pp. 823–826.
9. Kain, V., "Stress Corrosion Cracking (SCC) in Stainless Steels," In: *Stress Corrosion Cracking* (Cambridge, United Kingdom: Woodhead Publishing, 2011): p. 199.
10. Gooch, T.G., "Solidification Cracking of Austenitic Stainless Steel Welds," ASM Materials Week (Materials Park, OH: ASM International, 1990).
11. ASTM A262, "Standard Practices for Detecting Susceptibility to Intergranular Attack in Austenitic Stainless Steels" (West Conshohocken, PA: ASTM International).
12. Gooch, T.G., "The Unmixed Zone in CRA Weldments," *Stainless Steel World* (November 1995): p. 29.
13. Rogne, T., J.M. Drugli, T. Solem, H. Salbu, and H. Skjellevik, "Crevice Corrosion Properties of Weld Overlays of Ni-Based Alloys Compared to 6% Mo Steels for Seawater Applications," paper no. 696, Corrosion '98 (Houston, TX: NACE International, 1998).
14. Francis, R., *The Corrosion of Duplex Stainless Steels: A Practical Guide for Engineers* (Houston, TX: NACE International, 2018).
15. Gooch, T.G., "Corrosion Resistance of Welds in Duplex Stainless Steel," Duplex Stainless Steels '91 (Les Ulis Cedex A, France: Les editions de physique, 1991): p. 325.
16. ISO 17781 (latest edition), "Petroleum, Petrochemical and Natural Gas Industries–Test Methods for Quality Control of Microstructure of Ferritic/Austenitic (Duplex) Stainless Steels" (Geneva, Switzerland: International Standards Organization, 2017).
17. Ashland Chemical, "Resin Selection Guide for Chemical Resistance," London, United Kingdom, https://www.derakane.com (2020).
18. Clarke, S., and R. Francis, *The Selection of Materials for Mineral Processing Operations,* (Houston, TX: NACE International, 2016).

3 General Corrosion Problems

In this chapter, we address two types of corrosion common to all three types of desalination plants and that continue to cause numerous problems: galvanic corrosion and atmospheric corrosion. Corrosion issues relating to specific types of desalination plants are discussed in the next three chapters.

3.1 Galvanic Corrosion

In desalination plants, there is often a requirement to connect items together made of different metals. This can give rise to galvanic corrosion, which is also called "dissimilar metal corrosion," or "bimetallic corrosion." The causes of galvanic corrosion, the performance of different couples in a variety of environments, and the various methods of countering it are discussed in detail in Reference 1. Here, a basic introduction is given as it relates to desalination plants.

3.1.1 Galvanic Corrosion Fundamentals

Galvanic corrosion occurs when two dissimilar metals are connected in a conducting, corrosive medium, such as seawater. Each metal will develop an electrochemical potential (as measured against a standard reference electrode), and it is the potential difference between the two metals that drives the galvanic corrosion. The electropositive member (cathode) of the couple stimulates attack on the electronegative member (anode) of the couple, due to the potential difference between the two alloys. There have been instances of alloys with a potential difference of 800 mV being successfully coupled together, and others of severe galvanic corrosion with a potential difference of only 50 mV. Therefore, potential difference is a necessary but not sufficient condition for galvanic corrosion. The severity of the attack is controlled by the current density between the two metals and there are a number of factors that affect this, as discussed below.

In terms of galvanic series in seawater, Figures 3.1 and 3.2 show the typical potentials for some common alloys in seawater at 10°C and 40°C, respectively. With the development of alloys, and in relation to the desalination plants' conditions, Thierry et al.[2] determined the galvanic series for numerous alloys typically used in seawater under different conditions of temperature, oxygen, and chlorination. The effect of these environmental factors varies from alloy to alloy. For instance, temperature increase did not affect the potential of copper alloys, and chlorination did not affect the potential of a nickel alloy. Another important aspect to consider is that usually, reported potential values of corrosion resistant alloys in galvanic series are for passive states.

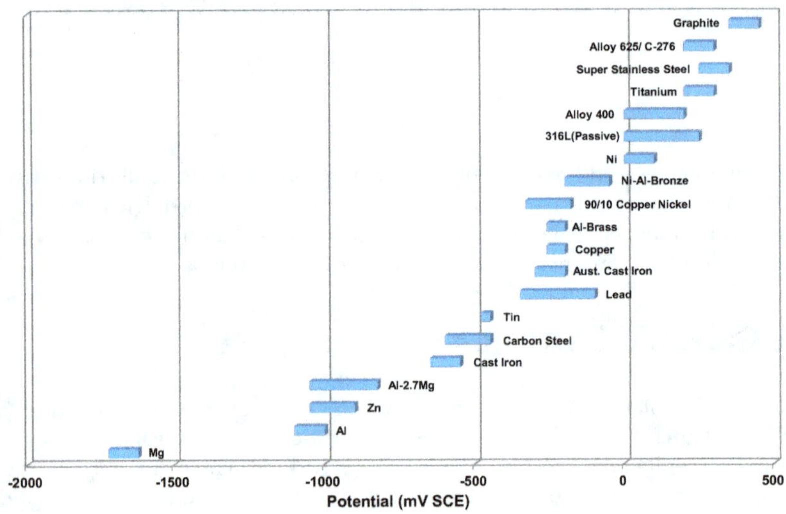

FIGURE 3.1 Galvanic series in seawater at 10°C. *Data courtesy of SINTEF and Roger Francis*

Most of the potential data show an apparent potential difference between alloys that can vary from ~20 mV to nearly 2 V. However, as described above, potential difference alone will not indicate whether there will be galvanic corrosion or not. Corrosion kinetics are controlled by the current density (Faraday's Law), so anything that affects the current density will influence galvanic corrosion. The two main factors that do this are area ratio and cathodic efficiency.

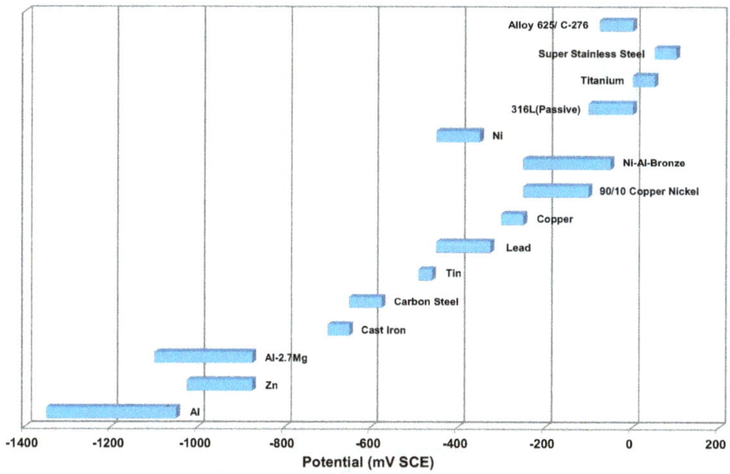

FIGURE 3.2 Galvanic series in seawater at 40°C. *Data courtesy of SINTEF and Roger Francis*

The larger the anode-to-cathode area ratio, the smaller will be the current density on the anode. This is successfully employed, for example, in nickel aluminum bronze (NAB) valves with nickel alloy trim. The large anode (valve body) to cathode (valve trim) area ratio means that the galvanic current density on the NAB is negligible from an engineering point of view.

3.1.2 Factors Affecting Galvanic Corrosion

If the metal forming the cathode is not very efficient at reducing dissolved oxygen, then the current density on the anode will not be too high, provided the anode is not smaller in area than the cathode. Typical factors that affect the cathodic efficiency include marine biofilms and calcareous deposits. The presence of a marine biofilm on the cathode has been observed to enhance oxygen reduction. This is the case for high alloy stainless steels and nickel-chromium-molybdenum (Ni-Cr-Mo) alloys where the marine biofilm in natural seawater forms within a short duration starting from 2 h, depending on the surface status (roughness) and the seawater characteristics (temperature, salinity, chlorination). This biofilm catalyzes the reduction of dissolved oxygen and means that lower alloy materials coupled to these alloys can suffer severe galvanic corrosion.[3] However, when the seawater is dosed with chlorine or hypochlorite, to give a residual chlorine of ~0.5 mg/L, the high alloy stainless steels and Ni-Cr-Mo alloys are much less efficient cathodes. An example of this is the coupling of 90/10 copper-nickel piping to high alloy stainless steel. In natural seawater, the corrosion of the 90/10 copper-nickel is greatly accelerated at the junction in natural seawater, but is hardly affected when the water is chlorinated, as shown in Figure 3.3. This is despite the potential difference being ~800 mV in chlorinated seawater.

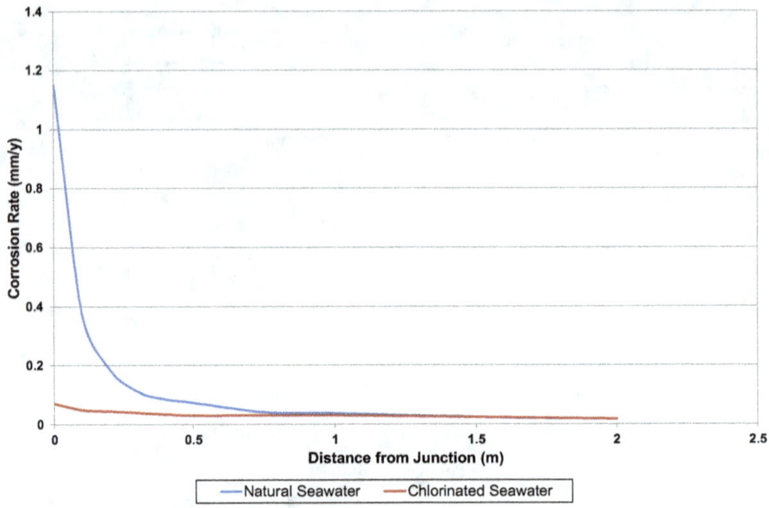

FIGURE 3.3 Corrosion rate versus distance from junction for 90/10 copper-nickel pipe coupled to a highly alloyed stainless steel pipe.[1]

The development of a calcareous deposit on the cathodic alloy in seawater is a situation enhanced by elevated temperature, as occurs in thermal desalination plants. This would limit the progress of the galvanic corrosion by playing the role of a physical barrier to oxygen reduction. Antiscaling chemicals are added to the seawater/brine in thermal desalination plants, but their effect is limited to delaying scale growth by disrupting its structure.

Graphite is a very efficient cathode with an electropositive potential, as shown in Figure 3.1. If graphite-loaded gaskets are used with flanged joints in a stainless steel chlorinated seawater system, galvanic corrosion can stimulate severe crevice corrosion of the stainless steel.[4] This has led to leakage of both 6%Mo austenitic and superduplex flange joints. Graphite-free gaskets should be used, such as synthetic fiber or neoprene.[4] For the same reason, glands on valves and pumps should use graphite-free packing. However, when spiral wound flange gaskets are used—mainly at higher pressures (>10 bar)—it is important that the metal used for the spiral windings is galvanically compatible with the piping alloy (Table 3.1). Graphite in spiral wound gaskets is acceptable provided it is never wetted.

Copper alloys and nickel-copper alloys (e.g., alloy 400) are also efficient cathodes in seawater and Francis points out that carbon steel coupled to copper-nickel suffers more attack than if coupled to the same area of titanium despite the greater potential difference with titanium.[1]

When it comes to welding, it is important to select a filler metal that is cathodic to the parent metal. This is because the area of the weld is very much smaller than that of the adjacent metal and will suffer very rapid attack if it is anodic to the parent metal. Severe corrosion has occurred with a carbon steel weld in seawater where the potential difference was only ~50 mV, but the large cathode-to-anode area ratio produced a high current density on the weld metal, causing the severe metal loss[1] (see Figure 2.1).

One other factor that is important in thermal desalination plants is aeration, because the lower the dissolved oxygen content, the lower is the current from the cathode. In the parts of MSF plants that operate at low oxygen contents (≤20 ppb), galvanic corrosion is not a major risk unless the anode-to-cathode area ratio is very small. However, where control of dissolved oxygen content is not good, and the concentration rises, not only does the corrosion rate of carbon steel increase, so does galvanic corrosion, such as at carbon steel–stainless steel junctions.

3.1.3 Galvanic Compatibility

There is a wealth of data from galvanic corrosion studies in seawater that show there are groups of metals that can safely be connected.[1] These are shown in Table 3.1, where alloys in the same group can be safely coupled together. The exception is in the group 4 alloys, where cast iron and carbon steel are in a separate category from aluminum alloys, as it is known that couples between steel and aluminum alloys will cause galvanic corrosion of the aluminum alloy. This is for ambient temperature seawater where, for instance, all the alloys in group 1 are passive. This means that high alloy stainless steel piping can be connected to titanium heat exchangers, provided there are no alloys in the heat exchanger from another group.

TABLE 3.1 Alloy groupings for galvanic compatibility in seawater at ambient temperature.

CATEGORY	TYPE	ALLOY
1	Noble; passive	Nickel-Chrome-Moly Alloys (Mo > 7%) 6%Mo Austenitic Stainless Steel Superduplex Stainless Steel Titanium and its alloys
2	Passive; not truly corrosion resistant	Alloy 400/K-500 904L 22%Cr Duplex Alloy 825 Alloy 20 316L
3	Moderate Corrosion Resistance	Copper Alloys Austenitic Cast Iron
4	Poor Corrosion Resistance	Carbon Steel Cast Iron
		Aluminum Alloys

From another perspective, the situation is less discussed when it comes to the risk of galvanic corrosion during acid cleaning operations, which aim to descale the heat exchangers in thermal desalination plants to restore the production and performance of the unit. In this case, the dominant cathodic reaction is hydrogen evolution due to the low pH (~2) of the acidified aerated seawater. However, this risk can be judged minor because of strict monitoring of acid cleaning operations associated with the use of a suitable corrosion inhibitor, as discussed by Meroufel.[5]

There are examples where mixing alloys from different groups can cause accelerated attack of the anodic metal. A good example is coupling NAB valves to superduplex stainless steel piping, where Francis showed that severe corrosion of NAB would occur in natural seawater and at a 10:1 cathode-to-anode area ratio in chlorinated seawater.[6] The service experience is that severe corrosion of NAB valves in superduplex piping systems occurred on both the Woodside Goodwyn A and BP Andrew platforms. The valves are usually thick wall castings so the corrosion manifests itself as the valve being unable to seal properly rather than as through-wall penetration. The seawater on Goodwyn A was chlorinated, but the seawater temperature varied from 20–40°C, i.e., higher than in Francis's tests.[6] The seawater on BP Andrew was unchlorinated at about 10°C, and severe corrosion occurred within one year. The main valves were replaced in superduplex stainless steel, and the others were changed as they failed. Chlorination was also retrofitted as it slows down the rate of attack (but does not stop it), because the stainless steel is a less efficient cathode, as discussed above.

Figure 3.4 shows a corroded NAB drain valve from the high pressure section of an SWRO plant with superduplex stainless steel seawater piping. The solution is to use a compatible drain valve material from Table 3.1, which is usually superduplex stainless steel, because of its lower cost.

FIGURE 3.4 Corroded NAB drain valve from a superduplex seawater piping system in the high pressure section of an SWRO plant.

3.1.4 Galvanic Corrosion Prevention

There are many ways to prevent or minimize galvanic corrosion. The main ones for seawater are discussed below, but there is more detail in Reference 1. For a new build, it is best to select alloys all from the same group to avoid problems. With desalination plants, this will usually either be group 1 or group 3 alloys. It is possible to use alloys from different groups in some instances, but make sure there is data to support their successful use together. With retrofits it can be more complicated, but selecting alloys from the same group generally avoids galvanic corrosion problems.

With piping systems, one method to prevent galvanic corrosion is the insulating gasket at a flanged joint. However, there are two potential problems that can make it ineffective. The first is that the insulation joint must be correctly assembled to ensure electrical isolation of the two pipes, and this requires skilled labor. The second is that the pipes are often connected externally, either through the electrical earthing system or via pipe hangers or supports, both of which will prevent the isolation flange working. It is important to check for electrical continuity between the two pipes before the system is filled with water, to ensure the isolation flange will work.

An alternative is to use an isolation spool pipe, which may be made of a polymer, such as FRP, or it can be a metal pipe of the cathode alloy with a high quality coating on the inside and up the flange faces. It must not be made of the anode piping alloy because there will be greatly accelerated attack at any small defects in the coating. For copper alloys coupled to stainless steel, the spool piece is usually 5–6 pipe diameters long. When isolating stainless steel from carbon steel the spool piece is typically 2.0–2.5 m long. Hence, there must be adequate space to fit these spool pieces and access to fit them.

In MSF flash chambers, there is sometimes a need to coat sensitive areas with stainless steel and it is important to avoid galvanic corrosion of carbon steel at the junction. One method is to coat the stainless steel up to about 25–50 mm from the junction, using a high quality coating, such as synthetic rubber, which will give an adequate life without introducing extra maintenance.

Another method of preventing galvanic corrosion is to use cathodic protection, usually as sacrificial anodes, because impressed current systems need to be very robust to work for long periods of time in desalination plants. An example is the increasing use of low alloy stainless steels for heat exchanger tube sheets with titanium tubes. Alloys such as 316L, 2101, 2304, and 2205 will suffer accelerated corrosion in seawater when coupled to titanium, as shown in Figure 3.5. Figure 3.5(a) shows the voluminous corrosion products between the tubes, while Figure 3.5(b) shows the deep pitting between the tubes after cleaning. This is normally prevented by inserting sacrificial low carbon steel anodes in each water box, but in this case, the anodes had accidentally been electrically isolated from the tube sheet and water box. It is important to use a low carbon ($<0.15\%$ C) steel so that fragmentation does not occur, and they corrode more or less uniformly. It is also important to size the anodes so that they will last between maintenance intervals for the plant without causing undue turbulence in the waterbox. It is common to size anodes to last for two years, when there will be less than 50% of the thickness left.

(A)

(B)

FIGURE 3.5 Appearance of 316L tube sheet with titanium tubes in the final heat exchanger in an MED plant: (A) as opened, (B) after cleaning, showing deep attack.

Where titanium or high alloy stainless steel tubes are retrofitted to a heat exchanger with copper alloy tube plates, there will be severe attack of the tube sheet unless it is cathodically protected, as shown in Figure 3.6. As for low alloy stainless steels, the tube sheets can be protected with low carbon steel anodes (C <0.15%), or pure iron. The use of zinc or aluminum alloy anodes should be avoided, because these will over-protect the tube sheets, and can cause significant scale buildup in the entrance to the tubes.[1]

FIGURE 3.6 Galvanic corrosion of NAB tube sheet with titanium tubes.

3.2 Atmospheric Corrosion

Desalination plants are generally situated close to the sea and in areas with lots of solar radiation. Marine atmospheres in warmer climate regions are very aggressive and it is important to use appropriate protective coatings on carbon steel, as well as selecting suitable grades of stainless steel for storage tanks, etc.

There are many environmental factors that affect atmospheric corrosion. The most important among them are humidity and temperature. Corrosion in direct contact with air at ambient temperature is generally not significant from an engineering point of view. For corrosion to occur, water must condense on the exposed metal surfaces or precipitate as rain. This will be followed by evaporation as the components dry out–this leads to the definition of a unified parameter for atmospheric corrosion, which is the time of wetness (sometimes abbreviated as TOW). Condensation is controlled by the relative humidity and the metal surface temperature. It is generally accepted that a relative humidity of 80% or more, plus a temperature greater than 0°C are required for significant condensation.[7] Rainfall, which has a washing effect, is dependent on the local weather and can vary greatly, even in a small region like the British Isles.

Corrosion in the atmosphere is generally not as great as that on fully immersed metal surfaces in water. As the water evaporates, any dissolved salts in it will concentrate. Hence, when these species are aggressive to a metal, the corrosion rate will increase as the water evaporates.

The principal cathodic reaction in atmospheric corrosion is usually the reduction of dissolved oxygen, and the reaction rates can be high, especially with thin water films. If the dissolved solids and gases content of the water is low, the conductivity will be low, reducing the rate of corrosion. As the films evaporate, however, the conductivity will increase because of the concentration effect.

It is important to be able to assess the corrosivity of atmospheres towards metallic corrosion and this is addressed in ISO 9223, which calculates a corrosion index based on the concentration of pollutants such as sulfur dioxide, chlorides, as well as time of wetness, temperature, etc.[8] In addition, the corrosion rate of some standard metals, such as steel, copper, zinc, and aluminum over a year can be factored in. This produces a ranking from C1 (mild) to C5 (very aggressive), and CX for extremely aggressive. ISO 9226 allows the atmospheric corrosivity assessment to be extended to areas beyond the scope of ISO 9223.[9]

Marine atmospheres close to the sea in warmer climates are usually classed as C5 or CX. At a distance from the sea (~0.5 km) the atmosphere may decrease to C4.

While this ranking works for many metals and alloys, it is not a good indicator of the performance of stainless steels in some atmospheres. This is because steel, copper, and zinc corrode uniformly in aggressive atmospheres, while stainless steels tend to suffer predominately localized attack (pitting). For example, an acidic atmosphere that is low in chlorides can be classified as aggressive for steel or zinc coupons, whereas stainless steels may suffer no attack. In marine atmospheres it is the chloride that is aggressive, and this may make some stainless steels pit. However, it will depend on other variables as to how aggressive a particular marine atmosphere is to stainless steel. The depth of localized corrosion attack on susceptible stainless steel grades in atmospheric environments can be quite shallow and may not affect the integrity of the component. In the following sections, some atmospheric corrosion data will be discussed for the concerned alloys together with outdoor degradation of polymeric materials.

3.2.1 Carbon and Low Alloy Steels

Figure 3.7 shows the extent of corrosion that can occur on steel in just a few years in a marine atmosphere if a suitable coating system is not used. With such an aggressive atmosphere it will be necessary to use a high quality coating system to protect carbon steel, similar to those used on offshore platforms. Such a system could be an inorganic zinc primer, two coats of epoxy, and a polyurethane topcoat to resist ultraviolet (UV) light. The types of coating for such duty and appropriate thicknesses for each layer are discussed in detail in References 10 and 11. It is advisable to consult a coating specialist when deciding on a suitable coating system for a specific item of equipment in a particular location.

FIGURE 3.7 Severe corrosion of carbon steel due to an inadequate painting system.

One particular application that can have serious repercussions is that of electrical enclosures. Figure 3.8 shows severe corrosion of a steel electrical enclosure at a Middle East desalination plant. FRP electrical enclosures with built-in UV protection are available off the shelf. Where the UV is particularly severe, the FRP enclosure can be given a coat of polyurethane paint to increase UV resistance and extend its life.

FIGURE 3.8 Badly corroded steel electrical enclosure.

3.2.2 Stainless Steels

There have been many atmospheric exposure tests of stainless steels, all of which have included a marine atmosphere and some have included semitropical or Middle East exposure.[12-16] A lot of the corrosion of 304L and 316L was cosmetic rust staining, but in a C5 or CX atmosphere 304L also showed pitting. Figure 3.9 shows a 304L chemical storage tank, which had pitted after a few years exposure in the Middle East (C5). In a CX atmosphere, even alloy 316L can pit and then duplex 2205 is required to give a good long term performance. The lean duplex stainless steels performed better than 304L, but not as well as 316L in an aggressive marine atmosphere.[14,16]

FIGURE 3.9 Corroded and pitted 304L stainless steel storage tank.

Hence, it is important to regularly check the condition of existing storage tanks made of stainless steel, particularly the depth of any pitting. When the pitting becomes significant it will be necessary to consider suitable repairs. First, the tank should be shot blasted (with nonsteel grit) to remove all rust staining. If the pitting is shallow (<0.5 mm) then the pits may be ground out. Nonsteel tools must be used to avoid steel contamination of the stainless steel and subsequent accelerated corrosion. If the pitting is deeper, then it will be safer to apply a protective coating; Addis et al. tested 10 systems for pitted 316L in the Gulf of Mexico.[17] They concluded that the only suitable system was an inorganic zinc primer, two coats of epoxy, and a urethane topcoat to resist UV. They also found that the pits were better covered if the primer was hand brushed on rather than sprayed.

3.2.3 Polymers

Most polymers will suffer degradation if exposed to strong sunlight with a high UV index, as occurs in tropical climates and the Middle East. The degradation manifests itself in the form of a chalky appearance or color change at the polymer surface. The effects of this can be minimized by taking some suitable precautions. The behavior of polymers under strong UV radiation and the prevention of degradation are discussed in Section 2.3.3.

References

1. Francis, R., *Galvanic Corrosion: A Practical Guide for Engineers*, 2nd ed. (Houston, TX: NACE International, 2017).
2. Thierry, D., C. Leballeur, andN. Larché, "Galvanic Series in Seawater as a Function of Temperature, Oxygen Content and Chlorination," *Corrosion* 74 (2017): pp. 147–152.
3. Dexter, S.C., and J.P. LaFontaine, "Effect of Natural Marine Biofilms on Galvanic Corrosion," *Corrosion* 54, 11 (1998): pp. 851–861.
4. Francis, R., and G. Byrne, "Gasket Selection for Stainless Steels in Seawater," *Materials Performance* 46, 10 (2007): p. 50.
5. Meroufel, A., "Corrosion Control During Acid Cleaning of Heat Exchangers," In: *Corrosion and Fouling Control in Desalination Industry* (Cham, Switzerland: Springer, 2020): p. 209.
6. Francis, R., "Coupling of Superduplex Stainless Steel and Cast Nickel–Aluminium Bronze in Seawater," *British Corrosion Journal* 34, 2 (1999): p. 139.
7. Mattsson, E., *Basic Corrosion Technology for Scientists and Engineers*, 2nd ed. (London, United Kingdom: IOM, 1996): p. 80.
8. ISO 9223 (latest edition), "Corrosion of Metals and Alloys-Corrosivity of Atmospheres—Classification, Determination and Estimation" (Geneva, Switzerland: International Standards Organization).
9. ISO 9226 (latest edition), "Corrosion of Metals and Alloys—Corrosivity of Atmospheres—Determination of Corrosion Rate of Standard Specimens for the Evaluation of Corrosivity" (Geneva, Switzerland: International Standards Organization).
10. Vincent, L.D., *The Marine Coating User's Handbook* (Houston, TX: NACE International, 2012).
11. Munger, C.G., and L.D. Vincent, *Corrosion Prevention by Protective Coatings*, 3rd ed. (Houston, TX: NACE International, 2014).
12. Fritz, J.D., J.F. Grubb, B.W. Parks, R.M. Kain, and S.J. Pikul, "The Influence of Iron Contamination on the Corrosion Resistance of Several Stainless Steels," paper no. 00457, Corrosion 2000 (Houston, TX: NACE International, 2000).
13. Lojewski, C., P. Boillot, and J. Peultier, "Atmospheric Corrosion Resistance of Duplex Stainless Steels: Results of a Field Exposure Program," Duplex 2010, Session IV.B.5 (Zutphen, Netherlands: KCI Publishing, 2010).
14. Le Manchet S., and D. Paul, "Atmospheric Corrosion Resistance of Stainless Steels for Architecture," paper no. 8880, Corrosion 2017 (Houston, TX: NACE International, 2017).
15. Mameng, S.H., R. Pettersson, L. Wegrelius, and C. Leygraf, "Effect of Surface Finish on the Atmospheric Corrosion of Duplex Grade UNS S32205; Results of a Field Exposure Program in Dubai," paper no. 7231, Corrosion 2016 (Houston, TX: NACE International, 2016).
16. Mameng, S.H., and L. Wegrelius, "Atmospheric Corrosion Resistance of Stainless Steel in the Middle East: Result of a Field Exposure Programme in Various Environments," paper no. 8964, Corrosion 2017 (Houston, TX: NACE International, 2017).

17. Addis, T.T., A. Bodington, C. Fonseca, and B.T.A. Chang, "Evaluation of Offshore Coating System on a Pitted Stainless Steel," paper no. 10487, Corrosion 2018 (Houston, TX: NACE International, 2018).

4 Multistage Flash (MSF) Distillation

In this section, the major components in an MSF plant will be discussed in terms of common alloys, corrosion risk factors and most frequent corrosion problems. In some cases, practical solutions are briefly considered but details are left for discussion in Chapter 8. Note that MSF brine recycle cross-tube process configuration is considered because of its wide application compared to other MSF configurations. The focus is paid to the internal corrosion in the desalting system because the corrosion in the seawater intake and the external ones are common to other coast-based industries using the seawater for different process purposes. While materials selected for this process have been widely discussed, the usage adopted in this guide is mostly taken from the reviews of Oldfield and Malik,[1, 2] supplemented by the experience of the authors. Common materials names are considered in the text and the corresponding Unified Numbering System (UNS) designation with typical chemical composition is given in the Appendix. For better visibility, each type of equipment has its own section, although there will be cross references between some of them when discussing common corrosion problems.

Figure 4.1 shows a schematic drawing of a typical MSF plant. During the 60 years of its existence around the world, the MSF process did not change in any revolutionary way from the design point of view except for the introduction of the ball cleaning system for units operated using high temperature antiscalants.[3] Then, with accumulation of experience from different MSF plants, the design lifetime increased from 25–40 years.[4]

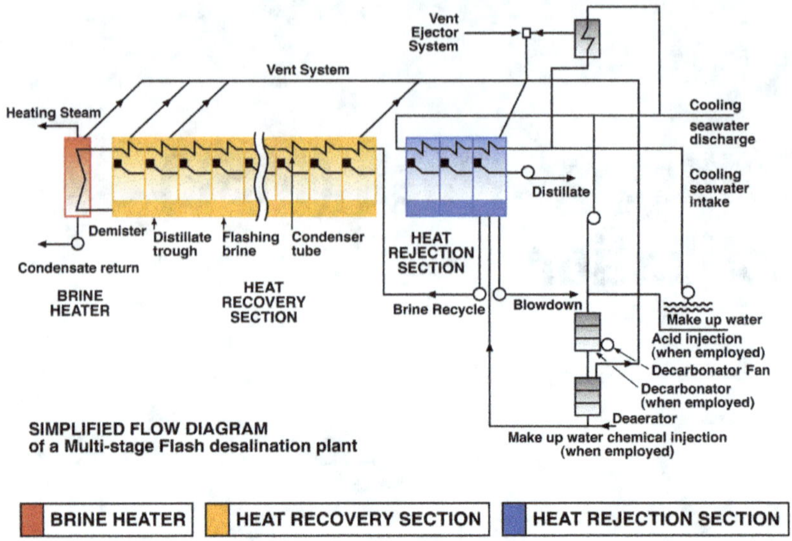

FIGURE 4.1 MSF Brine recycle desalination process.

4.1 Evaporators (Flash Chambers)

The terms "evaporator" and "flash chamber" are interchangeable, as observed in desalination topic publications. For more than 50 years, carbon steel dominated flash chamber construction materials except for the high temperature stages in the heat recovery section where it is clad with 316L or 317L stainless steels on the brine side. Stage dividers were made of painted carbon steel in early designs, then with corrosion experience, 316L stainless steel started to be the main choice.

The corrosion allowance of carbon steel MSF flash chambers considered in early MSF designed units was overestimated, reaching 12 mm. With operation and maintenance experience, this value was reduced significantly. Recently, duplex stainless steel flash chambers have been adopted, including the lean duplex 2304 (vapor side) and the standard duplex 2205 (brine side). This choice aims to decrease the life cycle cost of the unit through reduced maintenance compared to the externally painted and thick carbon steel.[5]

An important aspect to consider for ensuring the reliability of the flash chambers is the flushing procedure during shutdown and acid cleaning operation procedures. Stagnancy of saline or acidic media are detrimental to stainless steel, and proper flushing with distilled water followed by drying are recommended.

Water boxes for interstage linkage are also made of carbon steel clad internally by either nickel-aluminum-bronze (NAB, heat rejection) or 90/10 copper-nickel (90/10 Cu-Ni, heat recovery).

The demister is the important part that removes brine droplets, controlling the quality of the distillate. It is a wire pad placed on the upper part of each flash chamber and is made essentially from 316L stainless steel.

To ensure vacuum conditions inside the flash chambers for efficient brine flashing (and in the deaerator), and noncondensable gases removal, a venting system is used. It consists of a series of ejectors and condensers. The critical part of this system is the venting pipes installed on the top of the flash chambers with a cascade configuration that depends on temperature and CO_2 release, which varies from stage to stage, as studied by Glade.[6] Usually, these pipes are made of 316L stainless steel and in some cases of fiber reinforced plastic (FRP) composite for low temperature stages.[7] The FRP has usually been with epoxy resin and glass fiber reinforcement. The common materials and alternatives for the evaporators and the major internal parts are shown in Table 4.1.

The internal corrosion risks that are experienced in the MSF flash chambers during operation are affected by seawater pollution, oxygen, temperature, velocity, and noncondensable gases (O_2, CO_2). These factors play aggressive roles in both liquid (seawater, brine) and/or vapor phase exposures. The experiences show minor differences in the degree of corrosion attack between brine and vapor sides for the carbon steel. It has been noticed that oxygen ingress into the vapor space, especially in high temperature stages, enhances steel CO_2 corrosion by 10–40%.[8] Oxygen ingress has two sources, either from the flashed brine or air leakage from flanges. The vapor side corrosion is more critical on heat exchanger parts (tube, tube support) and will be discussed later.

TABLE 4.1. Standard material selection for the flash chambers and ancillary equipment in MSF plants. (SS = stainless steel, CS = carbon steel)

SECTION	COMPONENT	ALLOY	
		Most Common or Preferred	Alternative
Brine Side: High Temperature	Shell	CS	CS
	Lining or Coating	316L SS	90/10 Cu-Ni
	Floor	316L SS	90/10 Cu-Ni
	Side Walls	316L SS	90/10 Cu-Ni
	Interstage Walls	316L SS	90/10 Cu-Ni
	Roof	316L SS	90/10 Cu-Ni
	Weirs or Nozzles	316L SS	66/30/2/2 Cu-Ni-Fe-Mn
	Demister Pads	316L SS	Alloy 400
	Support Frames	316L SS	90/10 Cu-Ni

TABLE 4.1. (cont.) Standard material selection for the flash chambers and ancillary equipment in MSF plants. (SS = stainless steel, CS = carbon steel)

SECTION	COMPONENT	ALLOY	
		Most Common or Preferred	Alternative
Brine Side: Low Temperature	Shell	CS	CS
	Lining or Coating	Epoxy Coated CS	None
	Floor	Epoxy Coated CS	CS
	Side Walls	CS	Epoxy Coated CS
	Interstage Walls	CS	Epoxy Coated CS
	Roof	CS	Epoxy Coated CS
	Weirs or Nozzles	CS	Epoxy Coated CS
	Demister Pads	316L SS	Alloy 400
	Support Frames	316L SS	90/10 Cu-Ni
Vapor Side: High Temperature	Shell	CS	CS
	Lining or Coating	316L SS	None
	Floor	316L SS	CS
	Side Walls	316L SS	CS
	Interstage Walls	316L SS	90/10 Cu-Ni
	Roof	316L SS	CS
	Distillate Trays	316L SS	90/10 Cu-Ni
	Dist. Transfer Trough	316L SS	90/10 Cu-Ni
Vapor Side: Low Temperature	Shell	CS	CS
	Lining or Coating	316L SS	None
	Floor	CS	316L SS
	Side Walls	CS	316L SS
	Interstage Walls	316L SS	Epoxy Coated CS
	Roof	CS	316L SS
	Distillate Trays	316L SS	90/10 Cu-Ni
	Dist. Transfer Trough	316L SS	90/10 Cu-Ni

Special attention is paid to the oxygen and CO_2 removal by the deaeration of the makeup seawater and the venting process, respectively. The oxygen is kept below 20 ppb in the brine circulating in the heat recovery section by mechanical deaeration combined occasionally with the injection of oxygen scavenger (sodium sulfite: Na_2SO_3). Carbon dioxide is released from the thermal decomposition of bicarbonate ion, which is always present in seawater, and this will pass into the deaerated brine in plants using antiscalant additives. Most of the carbon dioxide is released in the high temperature stages where direct venting is considered. However, it is common practice to cascade incondensable gases from stage to stage and just vent them off in a few stages.

The removal efficiency of noncondensable gases depends on the number and location of venting points, as studied by some authors and observed in different MSF unit designs. Improper venting performance associated with bromine ingress were considered as an aggressive condition inducing stress corrosion cracking (SCC) of the 316L venting pipes, particularly for high temperature stages.[9] Typical SCC failure of 316L venting pipe can be seen in Figure 4.2. Venting pipes in 316L may also fail by SCC initiated from the external surface where wetting occurs (humid environment) and drying is caused by the hot internal gases, concentrating chlorides, and initiating SCC. This can be solved by replacing the 316L stainless steel with a more SCC resistant stainless steel grade, such as duplex 2205.

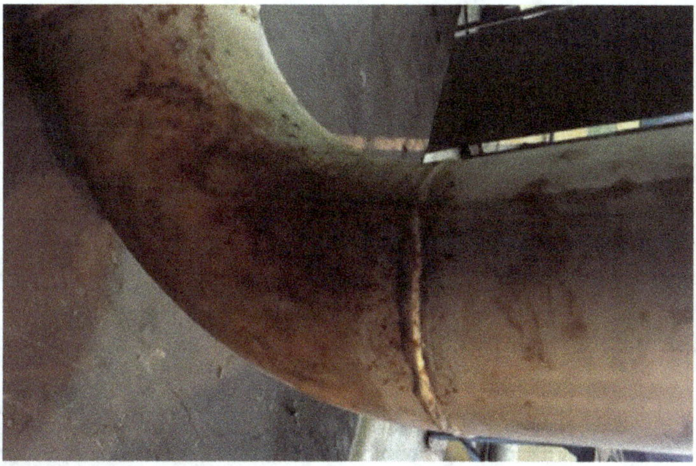

Part. Venting pipe.
Material. Stainless steel 316L
Corrosion form. SCC
Part age. 35 years

FIGURE 4.2 **SCC failure of 316L venting pipe.**
Photo courtesy of Saline Water Conversion Corporation

4.2 Heat Exchangers

The materials of construction for the main heat exchangers in an MSF plant are shown in Table 4.2. This includes both the most common and/or preferred alloy and other common alloys. While other alloys have been used for these heat exchangers, for example 90/10 copper-nickel in the heat rejection section, they are less common.

TABLE 4.2. Preferred materials for heat exchangers in MSF plants. (CS = carbon steel)

SECTION	COMPONENT	ALLOY	
		Most Common or Preferred	Alternative
Heat Rejection	Tubes	Titanium	66/30/2/2 Cu-Ni-Fe-Mn
	Anodes	Mild Steel	Mild Steel
	Tube Sheet	NAB	Naval Brass
	Water Boxes	90/10 Cu-Ni	Rubber lined CS
	Tube Supports	CS	316 SS
Heat Recovery (Low Temperature)	Tubes	90/10 Cu-Ni	Aluminum Brass
	Anodes	Mild Steel	9% Ni-Fe
	Tube Sheet	Naval Brass	90/10 Cu-Ni
	Water Boxes	90/10 Cu-Ni	Epoxy Coated CS
	Tube Supports	316 SS	Carbon Steel
Heat Recovery (High Temperature)	Tubes	90/10 Cu-Ni	70/30 Cu-Ni
	Anodes	Mild Steel	9% Ni Iron
	Tube Sheet	Naval Brass	90/10 Cu-Ni
	Water Boxes	90/10 Cu-Ni	90/10 Cu-Ni
	Tube Supports	316 SS	Carbon Steel
Brine Heater/ Dump Condenser	Tubes	66/30/2/2 Cu-Ni-Fe-Mn	90/10 Cu-Ni
	Anodes	Mild Steel	Mild steel
	Tube Sheet	90/10 Cu-Ni	Naval Brass
	Water Boxes	90/10 Cu-Ni	90/10 Cu-Ni
	Tube Supports	Carbon Steel	Carbon Steel
	Heater Shell	Carbon Steel	Carbon Steel

4.2.1 Heat Rejection

The seawater in this section is aerated, chlorinated (residual chlorine typically 0.25 mg/L) with a maximum temperature that can reach 40°C, and so all the usual corrosion problems that can occur with copper-nickel alloy heat exchanger tubes in aerated seawater may be seen.[10, 11] This is the reason that titanium is the most preferred alloy for the tubing because it is not susceptible to erosion corrosion or pitting, even in the case of polluted seawater.

In terms of tube sheets, carbon steel clad with NAB is usually used, associated with sacrificial anodes made of pure iron or mild steel ($C < 0.15\%$) to prevent galvanic corrosion between the noble titanium tubes and NAB tube sheet. The sacrificial anodes are frequently checked for proper electrical contact and replaced when consumed during annual shutdown for corrective maintenance.

The most common corrosion problems seen with copper-nickel tubes are pitting due to pollutants such as hydrogen sulfide (H_2S) or ammonia (NH_3). Where ammonia is present intermittently, the pitting can be prevented by ferrous sulfate ($FeSO_4$) dosing on first startup,[10] which aids in the formation of a protective film. If the plant has already been in service, it will be necessary to acid clean the tubes to remove the corrosion products before restarting with $FeSO_4$ dosing. The $FeSO_4$ dosing can be carried out for 60 days until a protective film has formed; it is not required after that. If the ammonia pollution is constant, then constant or regular $FeSO_4$ dosing may be required. The cost-effective alternative would be a retube with titanium grade 2 (Ti Gr2). The same practice can be followed when the pollutant is hydrogen sulfide. To assess and prevent pitting due to sulfides, Francis describes the use of a redox probe coupled with low level chlorine dosing.[10, 11]

Erosion corrosion is rarely a problem these days, because the safe design velocities for copper alloys are well known and well respected.[10, 11] Similarly, the lower limits should also be respected to prevent scale buildup that may lead to severe underdeposit or microbial corrosion. Where lodged obstructions are causing erosion corrosion, it is important to inspect and repair the filter system so that these objects do not enter.

4.2.2 Heat Recovery

The main corrosion problem for heat recovery heat exchangers is the vapor side corrosion affecting tubes and tube supports. Many authors reviewed the causes of this corrosion including carbon dioxide and oxygen.[12] Oldfield and Todd demonstrated that air leaking past gaskets and sight glasses in the heat recovery section could be sufficient to cause corrosion of carbon steel.[12]

For heat exchanger alloys (vapor side), the CO_2 buildup and its dissolution will acidify the condensate ($pH < 6$). In this range, and with reduced oxygen content, cuprous oxide is unstable depending on the potential, and copper alloys may suffer from corrosion that can be relatively severe at critical areas such as tube–tube support contact area. For copper-nickel tubes, it was observed that mainly the bottom part of the tube bundle suffers from this vapor side corrosion, as detected by eddy current inspection shown in Figure 4.3. This can be controlled by proper venting system design and oxygen ingress monitoring. A typical failure is shown in Figure 4.4 where the vibration of the tubes can induce localized attack, especially in the high temperature stages. Vibration of tubes can be due to both the high velocity of brine circulating inside tubes and/or the high velocity vapors, particularly in the high temperature stages. A galvanic couple has been proposed as an accelerating factor between the tube support plates made of 316L stainless steel (cathode) and the copper-nickel tubes (anode) in the presence of excess CO_2 or O_2 (Figure 4.4).

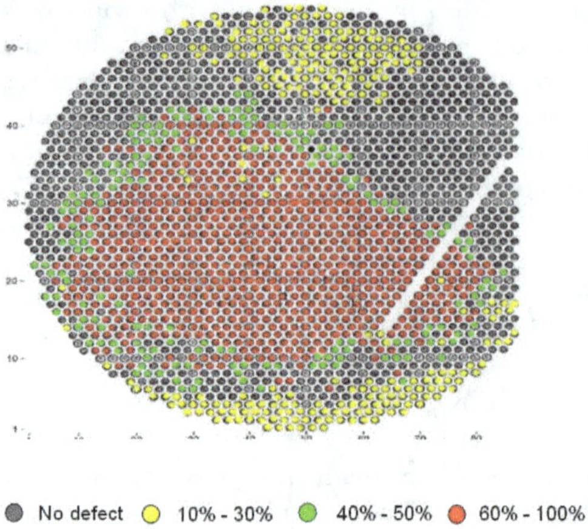

FIGURE 4.3 MSF first stage condenser tubes corrosion extent as revealed by eddy current after 14 years of service. *Photo courtesy of Saline Water Conversion Corporation*

<table>
<tr><td></td><td></td></tr>
<tr><td>Part. Heat recovery tube
Material. 90/10 Cu-Ni
Corrosion form. Pitting
Part age. 10 years</td><td>Part. Heat recovery tube support
Material. 316L
Corrosion form. Galvanic
Part age. 30 years</td></tr>
<tr><td colspan="2"></td></tr>
<tr><td colspan="2">Part. Dump condenser tube
Material. 66/30/2/2 Cu-Ni-Fe-Mn
Corrosion form. Erosion due to siphonic air release
Part age. 3 years</td></tr>
</table>

FIGURE 4.4 Typical failures of heat exchanger tubes in MSF distiller.
Photos courtesy of Saline Water Conversion Corporation and R. Francis

4.2.3 Brine Heater and Dump Condenser

The brine heater is the heart of the process, and it defines the flashing range and, therefore, the unit production capacity. The modified copper-nickel alloy with 2% iron and manganese added (66/30/2/2 Cu-Ni-Fe-Mn) is by far the most common alloy for the brine heater tubes and it has an excellent performance record.[1, 13] The brine heater water boxes and the tube sheets are usually made of 90/10 Cu-Ni. From the steam side, tubes are protected from the direct attack of steam flow by impingement baffles made of 304L stainless steel. Corrosion problems in the brine heaters are rare.

Dump condensers are often made of the same materials as the brine heater because they must cool the steam for recycling during necessary shutdowns. However, they usually use aerated seawater for cooling, and so can suffer the same corrosion problems with copper-nickel alloys as described in the heat rejection section.

Where sulfide pollution is the problem, it is often because the condenser is not used very often and it has been left full of stagnant seawater for an extended period. The sulfate reducing bacteria (SRB) in the seawater can become active and produce large quantities of corrosive H_2S. The solution is to keep the tubes dry when the dump condenser is not in use, or to maintain a low constant, chlorinated seawater flow through the condenser when not in use. A flow equivalent to a tube water velocity slightly above 0.5 m/s should be sufficient to avoid the buildup of deposits.[14]

Dump condensers are often high up so that they can use the siphon effect to increase water flows and reduce pumping demand. However, this can give rise to corrosion under some circumstances. Francis examined some 66/30/2/2 Cu-Ni-Fe-Mn tubes from the dump condenser in an MSF plant and observed severe erosion corrosion in some of the tubes, as shown in Figure 4.4.[13] All of the attack was near the outlet and toward the bottom of the condenser, which is typical of siphonic air release, as shown schematically in Figure 4.5. Gas is released under siphonic conditions, and the turbulence caused by this can result in severe erosion corrosion, even for an alloy like 66/30/2/2, which is very resistant to erosion corrosion.

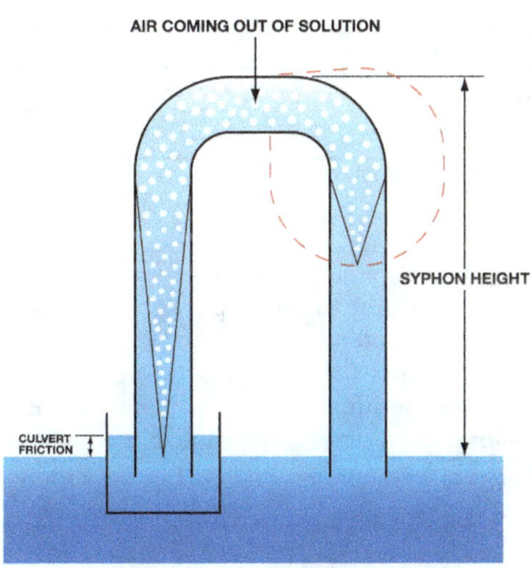

FIGURE 4.5 Schematic diagram showing release of air under siphonic conditions.

An analysis of condensers suffering from siphonic air release shows that the inlet and outlet sections of the second pass are most susceptible to attack.[15] Siphonic air release does not occur if the pressure in the tubing is maintained above 0.6 bar.[15] The solution was to adjust the valves in the seawater system to maintain an adequate pressure in the condenser tubes. An eddy current survey showed just a few tubes near the bottom were severely attacked and these were plugged. Since then, there have been no further problems.

4.3 Pumps

MSF desalination pumps are either horizontal (seawater handling) or vertical (brine handling) type and made usually from corrosion resistant alloys (CRAs). Table 4.3 summarizes typical pump materials for the different parts. Desalination pump components are subject to severe working conditions including aggressive environment and mechanical and/or hydraulic stresses. That is why some pumps are duplicated to ensure sustainable production. During operation, desalination pumps may be subject to relatively hot deaerated seawater (for make-up pump), or hot brine reaching 60,000 mg/L of total dissolved solids (TDS) (for brine recycle and brine blow-down pumps).

TABLE 4.3. Preferred materials for pumps for saline waters in MSF plants.

SECTION	COMPONENT	ALLOY	
		Most Common or Preferred	Alternative
Seawater	Casing	Superduplex	ACI*
	Impeller	Superduplex	CF8M/2205
	Column Pipe	ACI* (see text)	CF8M/2205
	Shaft	Superduplex	316L/2205
Brine	Casing	ACI*	NAB/2205
	Impeller	CF8M/2205	CF8M/2205
	Shaft	316L/2205	316L/2205

* Must be stress relieved.

Mainly 316L stainless steel was used for a long time before the recent recommendations to shift to the duplex grades. Alloy 316L is not resistant to crevice corrosion in seawater, but it worked because of galvanic protection from the austenitic cast iron (ACI) case. Precipitation hardening 17-7PH stainless steel and alloy 400 are also used for some parts (shaft sleeve, impeller ring) with satisfactory performance when they are galvanically protected by an ACI pump case.

ACI was used mostly for the pump column and bell mouth with questionable performance. In the last 20 years, there has been an increasing number of failures of ACI by SCC, mainly in the Middle East (Figure 4.6). The SCC failures have tended to be in the aerated seawater pumps, because corrosion is much less in the deaerated water sections. Failures occurred in 1–3 years with as-cast cases, and in 3–6 years with stress relieved cases. This was explained by residual stresses from the manufacturing process that were enough to initiate SCC when the alloy was in contact with warm seawater or hot brine.[16, 17] Then, a stress relieving post-treatment was recommended when using this material, which may raise the pump cost, or the use of duplex stainless steel, as decided by some desalination plants. Because the dissolved oxygen content in the brine is low, it is possible to use 2205 duplex for these pumps (Table 4.3).

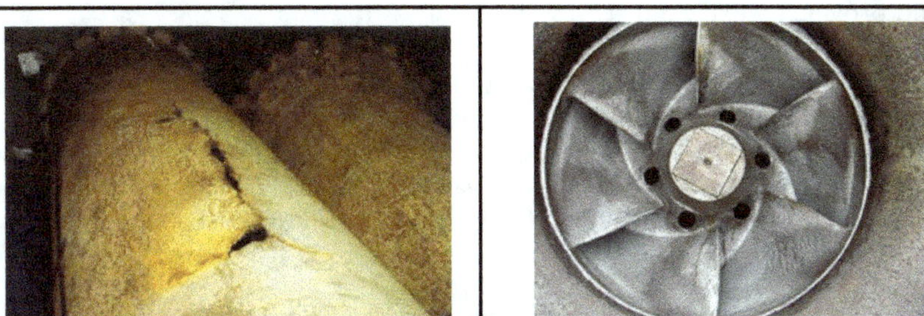

Part. Brine recycle pump column Material. ACI Corrosion form. SCC Part age. 10 years	Part. Brine recycle pump impeller Material. Duplex SS 2205 Corrosion form. Cavitation Part age. 30 years

FIGURE 4.6 Typical failures of brine recycle pumps in MSF desalination plants.
Photos courtesy of Saline Water Conversion Corporation and R. Francis

Where ACI pumps have been replaced, it is with superduplex stainless steel, because these grades (5A and 6A in ASTM A995) are resistant to crevice corrosion in aerated seawater[13] offering a good service experience. Superduplex stainless steel has been used for aerated seawater pumps for many years and it has generally been very successful. What problems have been seen were mostly due to poor quality castings. It is important that *all* duplex stainless steels are tested according to ISO 17781, to ensure that they will be adequate in service.[8] This standard covers appropriate tests for all grades of duplex stainless steel, as well as tests for weld qualifications (see Chapters 2 and 11).

Cavitation on the impeller is another chronic pump failure type that depends on the design and pump operating conditions. It occurs when the pumped liquid turns into vapor at low pressure generating destructive vacuum bubbles. This damage concerns almost all alloys including NAB (observed in early MSF days) and stainless steels as observed on duplex stainless steel as in the case of brine recycle pump at the suction side (Figure 4.6). The control of this type of damage is well understood by pump designers and desalination operators. Practically, pump engineers at the desalination plant should increase net positive suction available (NPSHa) to the pump or decrease the net positive suction required (NPSHr) by the pump.[18]

4.4 Valves

Valves in desalination plants play a crucial role, where their failure may cause unit shutdown. However, their investment remains minor (1–3%) compared to other systems, which implies some lack of consideration. All valves in contact with corrosive seawater or brine are butterfly-type, flanged with corresponding pipes, which should provide an expected service life of up to 15 years.

Typical valve materials include a cast iron or carbon steel body, with a rubber lining to prevent corrosion, while the disk is usually cast NAB. For the shaft, alloys 400 or K-500 are often used, although NAB and other alloys are sometimes used. A typical failure observed on seawater valves is corrosion developing under the rubber coated steel body as shown in Figure 4.7. This was caused by the ingress of seawater on the damaged lining area that was improperly repaired.

Part. Cooling water valve Material. Rubber coated carbon steel Corrosion form. Under rubber corrosion Part age. 17 years	Part. Butterfly valve disk (part) Material. NAB Corrosion form. Pitting due to sulfides Part age. 3 years

FIGURE 4.7 Typical failures of MSF desalination control valves.
Photo courtesy of Saline Water Conversion Corporation

NAB may suffer from severe corrosion due to seawater pollution by sulfides, as observed for a seawater valve shown in Figure 4.7. Following this type of corrosion risk, the introduction of stainless steels for disks and shafts gained more acceptance in the market because of the successful field performance. Superduplex disks have performed well.

Superduplex stainless steels can suffer from corrosion in case of poor casting methodology, which causes "worm holes" and other surface impurities leading to localized disk corrosion. This can be avoided by insisting on dye penetrant testing of all castings and removal of small defects by grinding. Larger defects may require grinding and weld repair. Major weld repair (as defined by ASTM A995) will require postweld heat treatment. Proper pickling is essential to remove impurities that may lead to corrosion in service. Disks and sealing points should have a good finish to enhance the resistance to pitting corrosion. A sealing area that appeared less well-finished corroded much faster than others with a highly polished finish.

In the case of brine where the oxygen concentration is negligible, pitting corrosion may initiate at the shaft's primary sealing point causing the penetration of salt water to the shaft area and behind the replaceable liner.

A recent solution to the duplex corrosion failure was proposed, which consists of coating the valve disk with a tough ultra-high molecular weight polyethylene (UHMWPE) layer.[19] This solution showed excellent performance in flue gas desulphurization (FGD) plants where extremely abrasive and corrosive conditions are encountered (chloride concentrations up to 150,000 mg/L and average temperatures of 50–60°C).

References

1. Oldfield, J.W., "Survey of Material Usage in MSF Plants over the Past 25 Years," IDA World Conference (Topsfield, MA: International Desalination Association, 2003).
2. Malik, A.U., and S. Al-Fozan, "Corrosion and Materials Selection in MSF Desalination Plants," *Corrosion Review* 229 (2011): pp. 153–175.
3. Morris, R.M., "The Development of the Multi-Stage Flash Distillation Process: A Designer's Viewpoint," *Desalination* 93 (1993): pp. 57–68.
4. Meroufel, A., "Corrosion in Thermal Desalination Plants," In: *Corrosion and Fouling in Desalination Industry* (New York, NY: Springer, 2020).
5. Olsson, J.O., V. Jagerstrom, and I. Resini, "MSF Flash Chambers of Solid Duplex Stainless Steel," IDA World Congress (Topsfield, MA: International Desalination Association, 2003).
6. Glade, H., and K. Genthner, "Measurements of CO_2 release rates in individual stages of MSF distillers," IDA World Congress (Topsfield, MA: International Desalination Association, 1999).
7. Al-Fozan, S., and A.U. Malik, "MSF Evaporators Flash Chamber Materials Evaluation after 20 Years in Service," IDA World Congress, MP07-159 (Topsfield, MA: International Desalination Association, 2007).
8. Skaperdas, G.T., and H.H. Uhlig, "Corrosion of Steel by Dissolved Carbon Dioxide and Oxygen," *Industrial & Engineering Chemistry* 34 (1942): pp. 748–754.
9. Lee, W.S.W., J.W. Oldfield, and B. Todd, "Corrosion Problems Caused by Bromine Formation in Additive Dosed MSF Desalination Plants," *Desalination* 44 (1983): pp. 209–221.

10. Francis, R., *The Corrosion of Copper and Its Alloys: A Practical Guide for Engineers* (Houston, TX: NACE International, 2010).
11. Francis, R., "Copper Alloys in Seawater: Avoidance of Corrosion," CDA Publication 225, Hemel Hempstead, United Kingdom: Copper Development Association (2016), https://www.copper.org/applications/marine/cuni/pdf/pub-225-copper-alloys-in-seawater-avoidance-of-corrosion.pdf.
12. Oldfield, J.W., and B. Todd, "Vapour Side Corrosion in MSF Plants," *Desalination* 66 (1987): pp. 171–184.
13. Francis, R., *Avoiding Corrosion in Desalination Plants* (Houston, TX: NACE International, 2019).
14. Lefebvre, Y., "Materials Used in Seawater Systems," In: *Seawater Circuits: Treatments and Materials* (Paris, France: Éditions Technip, 1998): p. 181.
15. Page, G.G., "Influence of Syphonic Conditions on Erosion-Corrosion of Brass Heat Exchanger Tubes: A Review," *New Zealand Journal of Science* 26 (1983): p. 415.
16. Malik, A.U., S. Basu, I. N. Andijani, N. A. Siddiqi, and S. Ahmad, "Corrosion of Ni-resist Cast Irons in Seawater," *British Corrosion Journal* 28 (1993): pp. 209–216.
17. Francis, R., *Selection of Materials for Seawater Cooling Systems–A Practical Guide for Engineers* (Houston, TX: NACE International, 2010).
18. Xylem Applied Water Systems, "Pump Cavitation and How to Avoid It: Best Practices in Pump System Design" (2015), https://www.xylem.com/en-us/support/case-studies-white-papers/pump-cavitation-and-how-to-avoid-it/.
19. Darpin, C., and R. Davia, "Cause and Effect Chain of Valve Technology Applied in Desalination Plants," IDA World Congress (Topsfield, MA: International Desalination Association, 2011).

5 Multiple Effect Distillation (MED)

In comparison to multistage flash (MSF) distillation, MED is a simple process with a reduced record of corrosion failures. Process wise, vapors supplied from a boiler or generated from seawater evaporation condense inside heat exchanger tubes to produce fresh water. The process has the advantage that it can come in a wide range of sizes with a consecutive number of effects operating at reduced pressure. MED plants, like MSF plants, are cost effective when installed in dual purpose plants together with a steam and power co-generation plant. The source of heat can be low-pressure steam or hot water (MED only). In most MED plants, a thermo-vapor compressor (TVC), or a mechanical vapor compressor (MVC) is added to improve thermal efficiency. These are known as thermo-compression distillation (TCD) and MVC plants, respectively. The higher efficiency of TCD plants means that they are the most common variant. Figure 5.1 shows a schematic drawing of a typical TVC-MED plant, although it may contain from 2–14 stages.

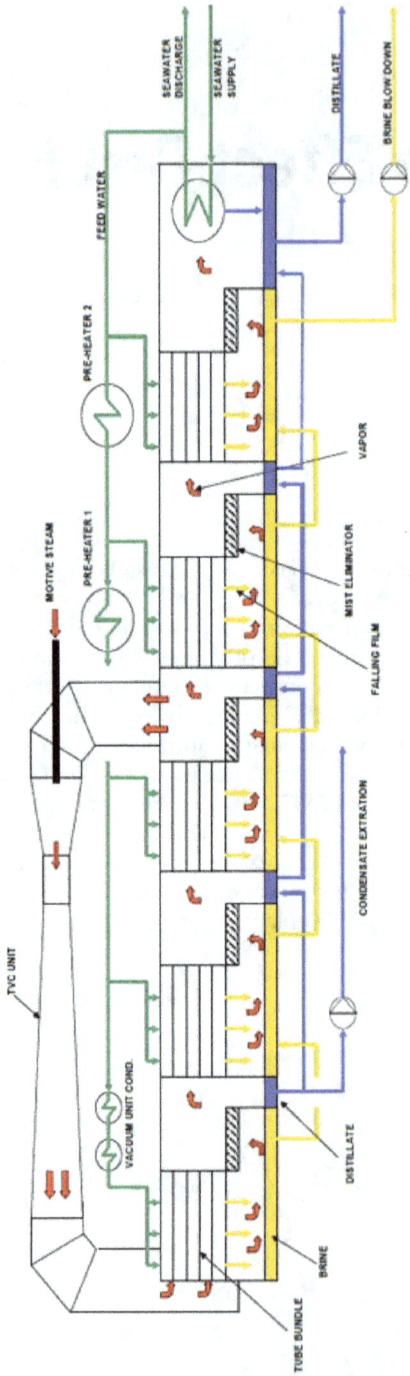

FIGURE 5.1 Schematic diagram of a typical MED plant.

Many corrosion issues common with the MSF distillation process will not be covered in detail in this chapter, such as stress corrosion cracking of seawater pumps and pitting corrosion of nickel-aluminum-bronze (NAB) valve parts. These problems and related recommendations are discussed in Chapter 4.

The following sections describe the materials often used in different parts of MED plants and the common corrosion problems seen there.

5.1 Evaporators

Table 5.1 shows the preferred alloys for the evaporators, as well as other alloys used in the past. The compositions of all the alloys are shown in the appendix of this book.

TABLE 5.1 Materials used in MED evaporators.

COMPONENT	ALLOY	
	Common or Preferred	Other
Shell (cylindrical)	Lean duplex	316L
Shell (rectangular)	Lean duplex and 2205 base	316L
Tubes	Titanium	Al-brass or 90/10 Cu-Ni
Tube Sheets	Lean duplex	316L
Support Plates	Lean duplex	316L

The evaporator shells, baffles, tube sheets, and tube support plates were traditionally in 316L stainless steel. However, since 2000 the cost of duplex stainless steels compared to austenitic alloys has dropped because they contain less nickel and molybdenum. Currently, lean duplex is lower in cost than 316L and cylindrical evaporators are now made of lean duplex, which has a similar corrosion resistance, but greater resistance to stress corrosion cracking (SCC), and higher strength. Section thicknesses can then be reduced–within design code restraints–to further lower costs.[1] Because of the existence of two environment, i.e., brine and vapor, two duplex stainless-steel grades are usually specified: duplex 2205 (brine side) and lean duplex 2304 (vapor side).

The evaporator tubes were traditionally in copper alloys such as aluminum brass or 90/10 copper-nickel (Cu-Ni), except for the top three rows, which were in titanium. The use of titanium here was to prevent erosion-corrosion from the spray heads at the top of the evaporators. Erosion has been observed where three full rows of titanium were not fitted at the top. A full row is two offset rows so that water will encounter at least three titanium tubes before the aluminum brass. Erosion has been seen on aluminum brass tubes due to excess buildup of silt in the main seawater sump or basin.[1] It is important that such sumps are dredged at regular intervals to avoid the buildup of silt. Typical aluminum brass failures in the form of erosion-corrosion for a velocity that exceeds 1.8 m/s were recorded (Figure 5.2).

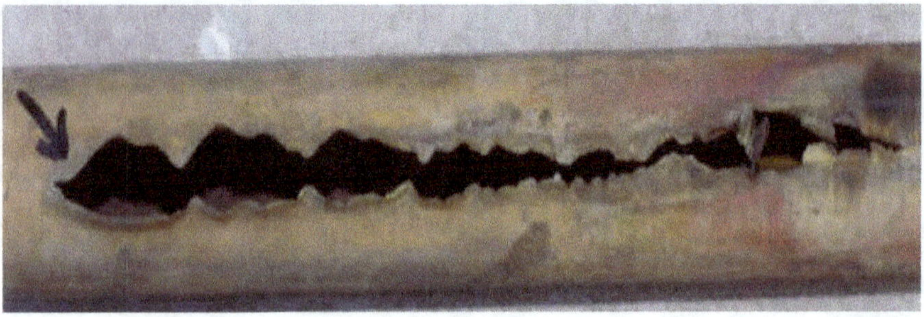

FIGURE 5.2 External erosion-corrosion of MED Al-brass tube.

However, the trend is now to use all titanium tubing in the evaporators because the cost of titanium has dropped to little more than that of aluminum brass. This removes the risk of corrosion by the seawater, but it is necessary to provide extra support plates, compared with aluminum brass tubes, because of the reduced fatigue resistance of titanium compared with brass. Where it is a new design, the extra support plates can be included easily, but a re-tube in titanium from aluminum brass will mean the provision of supports between the tubes (e.g., neoprene) to reduce vibration and the risk of a fatigue failure, if the fitting of extra tube support plates is to be avoided. There is no risk of galvanic corrosion of the lean duplex with titanium tubes in the evaporators because the seawater has a low dissolved oxygen and is treated to prevent the formation of a biofilm. Under these conditions, the electrochemical potentials of the two alloys are very similar and titanium is a poor cathode under such conditions.[2]

Other tubing materials have been suggested, although they have not yet been included in MED specifications. These include thin superferritic stainless steel (compositions in appendix), and conductive polymer composites.[3,4] Superferritic stainless steels were suggested based on their thermal conductivity performance observed in power plants. However, their sensitivity to crevice corrosion was not confirmed despite some laboratory tests. For conductive polymer composites, some preliminary pilot tests were successful, but their long-term behavior and their sustainability after chemical descaling operations remain unknown.

5.2 Final Condenser

This is a conventional heat exchanger with vapor on the outside of the tubes, and aerated, chlorinated seawater at ambient temperature inside. The materials of construction are shown in Table 5.2.

TABLE 5.2 Materials used in MED final condensers.

COMPONENT	ALLOY	
	Common or Preferred	Other
Tubes	Titanium	Titanium and 90/10 Cu-Ni
Tube Sheets	2205	316L
Water Boxes	2205	316L and Neoprene lined steel
Anodes	Mild steel	Mild steel
Shell	Carbon steel/lean duplex	Carbon steel
Support Plates	Lean duplex	316L or Carbon steel

Traditionally, the tubes were made of one of the copper-nickel alloys, but with the reduced price and their virtual immunity to the corrosion problems that can affect copper alloys (erosion-corrosion, sulfide attack etc.[5]) titanium tubes have become more or less standard. To avoid fatigue failure problems, it is necessary to provide additional tube support plates compared with copper alloy tubes, as discussed previously (Section 5.1).

One of the less common corrosion failure modes in the MED final condenser is microbiologically influenced corrosion (MIC). A typical case occurred with modified 70/30 Cu-Ni in contact with cooling water from the Red Sea at 35°C. There was a large amount of pitting at the 6 o'clock position after only six years of operation (Figure 5.3). Meroufel et al.[6] discussed the root cause of this failure, which was a seawater velocity below the limit to prevent a buildup of deposits, which allowed anaerobic microbial activity. The low velocity was adopted during a low production period based on reduced water demand.

FIGURE 5.3 MIC on MED final condenser copper-nickel tubes after six years of operation.

The tube sheets used to be copper-nickel and then 316L stainless steel, with sacrificial mild steel anodes in the water boxes to prevent galvanic corrosion of the 316L. Nowadays, it is usual to use 2205 duplex stainless steel for the tube sheets, which can be of reduced thickness compared with 316L, because of the higher strength of 2205. Sacrificial mild steel anodes are still required with 2205 tube sheets, to prevent galvanic corrosion. The anodes should be mild steel ($C < 0.15\%$)

to avoid fragmentation issues when they are corroding. They should be suitably sized to last one or two years before they need replacement. In addition, the water box should be suitably designed so that the incorporation of the anodes does not affect the water distribution or introduce undue turbulence.

The water boxes used to be carbon steel lined with 90/10 copper-nickel, or were rubber lined (e.g., neoprene), but more recently they have been fabricated from 316L and the anode size was increased so that both the water box and tube sheet were protected. With the reduced cost of duplex stainless steels, the water boxes are now mostly made of 2205 duplex with protection from mild steel anodes. The higher strength of 2205 compared with 316L means that wall thicknesses can be reduced to lower costs.

On the vapor side, the shell, baffles, and tube support plates were commonly made of 316L stainless steel, although 304L was sometimes used for the baffles and tube support plates. With the reduced cost of duplex alloys, as discussed above, the shell, baffles, and tube support plates are now made of lean duplex stainless steel.

5.3 Pumps

Depending on the design, an MED plant may have separate seawater lift and recirculation pumps, or the seawater may be lifted and circulated by the same pump. The common materials of construction are shown in Table 5.3.

TABLE 5.3 Materials used for pumps in an MED plant.

SECTION	COMPONENT	ALLOY	
		Common or Preferred	Alternative
Seawater Supply	Case	Superduplex	ACI*
	Impeller	Superduplex	CF8M or 2205
	Shaft	Superduplex	316L or 2205
	Column Pipes	Superduplex	ACI*
Seawater Recirculation	Case	Superduplex	ACI*
	Impeller	Superduplex	CF8M or 2205
	Shaft	Superduplex	316L or 2205
Brine Blowdown	Case	ACI+	NAB
	Impeller	CF8M/2205	CF8M or 2205
	Shaft	316L/2205	316L or 2205
Produced Water	Case	CF8M	NAB
	Impeller	CF8M	CF8M
	Shaft	316L	316L

* See text
\+ Must be stress relieved

The seawater supply and circulation pumps both handle warm, aerated seawater and, hence, tend to use the same materials. Austenitic cast iron (ACI) has traditionally been used for pump cases and column pipes, with stainless steel impellers and shafts. However, in the last 20 years there has been an increasing number of failures of ACI by SCC, mainly in the Middle East (see Figure 4.6 in the MSF section, Chapter 4). The SCC failures have tended to be in the aerated seawater pumps, because corrosion is considerably less in the low oxygen water sections. Failures occurred in 1–3 years with as-cast pump cases, and in 3–6 years with stress relieved cases.

In terms of level of stress, Zhou and Turnbull[7] carried out unpublished research into SCC of ACI and found that the threshold stress for SCC of ACI is very low—not much greater than the residual stresses after stress relieving. After a few years, corrosion proceeded along the boundary between the chromium carbide particles and the matrix.[8] These narrow fingers acted as stress raisers and acted as initiation sites for SCC. Older pumps did not appear to have this problem, and it can be said that this is due to the thicker sections used in older pumps. With the increasing price of nickel, there has been a trend to reduce pump case thickness, thereby increasing the stress on the case.[8]

Where there have been repeated failures of ACI pump cases, the cases have been replaced with superduplex, either just the case itself or as a whole new superduplex pump. Wherever superduplex is just used for a replacement case, note that any low alloy (PREN < 40) stainless steels elsewhere in the pump will no longer be galvanically protected and may suffer crevice corrosion. For more information about materials for seawater pumps see the relevant section in Chapter 4.

For the brine blowdown pumps, the dissolved oxygen content is generally low, so that corrosion rates are low and the risk of SCC of ACI is reduced. Hence, failures of ACI are less common than in the aerated seawater system, and stress relieving of the castings seems to be effective in preventing SCC in this environment. Hence, the materials for the brine blowdown pumps tend to have ACI cases and stainless steel internals. Because of the low oxygen content, grade 4A (cast 2205 duplex) could also be considered for the blowdown pump case. Cathodic protection was also proposed to overcome the SCC risk with ACI.[9]

5.4 Piping

Like MSF plants, the main seawater piping tends to be fiberglass reinforced plastic (FRP), particularly for larger diameter pipes that have both buried and aboveground sections. These pipes are sensitive to the operating conditions, and pressure must be well controlled to avoid water hammer or transient events that may led to the sudden burst or collapse affecting the whole desalination process.

Older plants may use coated carbon steel or concrete for the large intake pipes. For the brine piping, the low oxygen content means that 316L stainless steel is common. Lean duplex would also be acceptable, but at the time of writing, lean duplex alloys were not readily available in pipes, fittings, and flanges.

5.5 Valves

Most of the valves in an MED plant are on-off with two dominant models: ball and butterfly valves as the most cost-effective choices. In the seawater system, these tend to be cast steel or cast-iron bodies with a good quality internal coating of neoprene or another polymer. The disk or ball is often NAB and this generally works well. There are two problems, though: pitting in polluted seawater and preferential phase corrosion due to improper material quality, which were investigated by Francis.[5, 7] The former problem is discussed in Chapter 4 (valves section). Francis[1, 5] has reported selective phase corrosion of NAB disks when the final heat treatment discussed in Chapter 2 (Section 2.2.3) was not carried out.

The valves in the brine lines are often made of 316L or NAB, and the same comments apply about a final heat treatment for NAB, as for seawater valves (Chapter 2, Section 2.2.3.

The most common material for valves in contact with produced water is 316L (or the cast version CF8M), which rarely gives problems. Polymer valves may be used in polymer lines.

5.6 Venting System

In larger MED plants, typically more than 1,000 m^3/d, the vent system is a steam ejector-based system, usually in two stages, and the vent gases are cascaded down to the final condenser. There they are drawn into the first stage ejector and then to an intercondenser to remove most of the vapor, before going to the second ejector, where the vapor is condensed in the after condenser. Temperatures at the ejector discharge stages are high, being around 120–130°C after the first stage and around 140–160°C for the second stage. This is similar to an MSF system, but without a precondenser, whose duty is served in an MED plant by the final condenser.

Materials are typically 316L as a minimum, but where corrosion is occurring this could be upgraded to 2205, 904L, 6%Mo austenitic or superduplex stainless steel, depending on price and availability. At the time of writing, duplex stainless steels are significantly cheaper than austenitic ones.

The condensers tend to be tubed in titanium as is done for the final condenser, to avoid corrosion problems. Because these are relatively small units compared with the final condenser there is no significant cost penalty. The other components in the condensers are usually the same as described above for the final condenser.

Where temperatures exceed 300°C in the motive steam design, then duplex stainless steels are not suitable, because alpha prime phase will precipitate and greatly reduce ductility. In this case the materials must be austenitic stainless steel or titanium.

Problems have occurred with 316L stainless steel in vent systems, and SCC may occur anywhere that water with chlorides collects, and the temperature exceeds 60°C. This means that cracking may originate from the outside or the inside. A switch to 2205 duplex stainless steel will usually solve this because of the alloy's excellent resistance to SCC. Two critical parameters, i.e., temperature and chlorides, are shown in Figure 5.4, to indicate the limits of use for some stainless steels at near neutral pH. Such guides are widely used in many industries to avoid SCC problems.

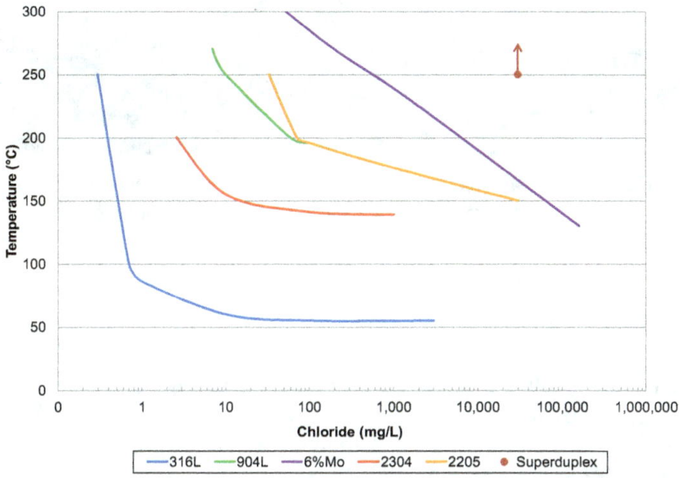

FIGURE 5.4 SCC limits for some stainless steels at near neutral pH.

5.7 Condensate Heater

Considering the construction materials of the condensate heater, a combination of stainless steel as tube and carbon steel as tube sheet is typical. This means that the condensate fluid should be oxygen free and have very low conductivity. Otherwise, an inevitable galvanic corrosion cell, accelerated by temperature, would happen, as occurred in the failure case shown in Figure 5.5. This was attributed to air ingress into the sodium sulfite tank (added as an oxygen scavenger), which increased the oxygen concentration in the water entering the condensate heater. Then, severe corrosion was discovered on the cap of the condensate heater. In the absence of oxygen monitoring, the migration of corrosion product up to the vent condenser caused a blockage at its inlet, which then caused severe scale formation on the shell side of the steam transformer. The consequence of these interrelated issues caused a decrease in the steam required for the MED process, which showed as a decline in the distilled water production.

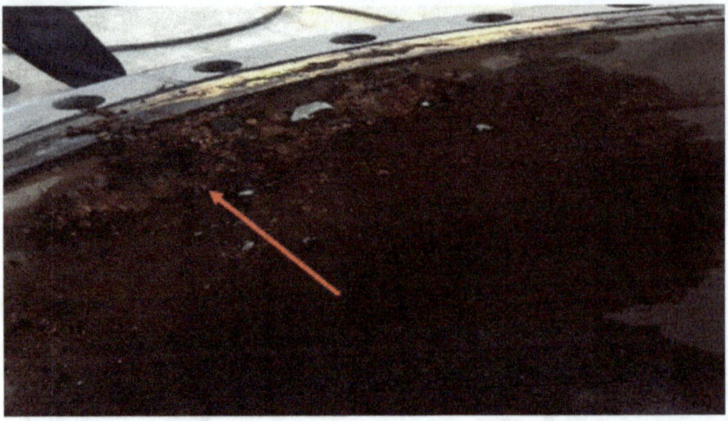

FIGURE 5.5 Corrosion product observed on the condensate heater cap.
Photo courtesy of Saline Water Conversion Corporation

References

1. Francis, R., *Avoiding Corrosion in Desalination Plants* (Houston, TX: NACE International, 2010).
2. Francis, R., *Galvanic Corrosion: A Practical Guide for Engineers*, 2nd ed. (Houston, TX: NACE International, 2017).
3. Peultier, J., V. Baudu, P. Boillot, and J.C. Gagnepain, "New Trends in Selection of Metallic Material for Desalination Industry," paper no. IDAWC/DB09, IDA World Congress (Topsfield, MA: International Desalination Association, 2009).
4. Orth, T., D.-C. Alarcón-Padilla, G. Zaragoza, P. Palenzuela, and P. Mueller, "Shaping the Future of Thermal Desalination with High Performance Heat Conducting Polymer Tubes," IDA paper no. IDAWC/58054, IDA World Congress (Topsfield, MA: International Desalination Association, 2017).
5. Francis, R., *The Corrosion of Copper and Its Alloys: A Practical Guide for Engineers* (Houston, TX: NACE International, 2010).
6. Meroufel, A., A. Al-Sahari, M. Ayashi, and A. Alenazi, "Microbial Corrosion in Water Desalination Condenser Cupronickel Tubes," *Materials Performance* 57 (2018): pp. 15–17.
7. Zhou, S., and A. Turnbull, "Stress Corrosion Cracking of Austenitic Cast Iron in Synthetic Seawater," NPL Report MATC(D)32 (London, United Kingdom: National Physical Laboratory, July 2001).
8. Francis, R., *The Selection of Materials for Seawater Cooling Systems, A Practical Guide for Engineers* (Houston, TX: NACE International, 2006).
9. Miyasaka, M., and N. Ogure, "Stress Corrosion Cracking of Austenitic Cast Irons in Seawater and Brine and its Prevention," *Corrosion* 43 (1987): pp. 582–588.

6 Seawater Reverse Osmosis (SWRO) Desalination

The reverse osmosis (RO) desalination process continues to be adopted widely in the recent desalination plants installed globally and seawater is the most common feed water. The principle of this process is to desalt the seawater by pushing pretreated seawater at high pressure (typically 60–70 bar [6–7 MPa]) through a semipermeable membrane to obtain a distillate that is post-treated to meet recommended drinking water quality (Figure 6.1). A typical RO desalination plant consists of three major systems: seawater pretreatment, high pressure, and post-treatment. Corrosion and material selection for this process directly affect the water production in terms of cost, quantity, and quality. The effects of these can be assessed through the maintenance and capital costs and membrane reliability. In this chapter, corrosion risks and guidelines for their mitigation will be limited to the pretreatment, high pressure, and brine reject sections.

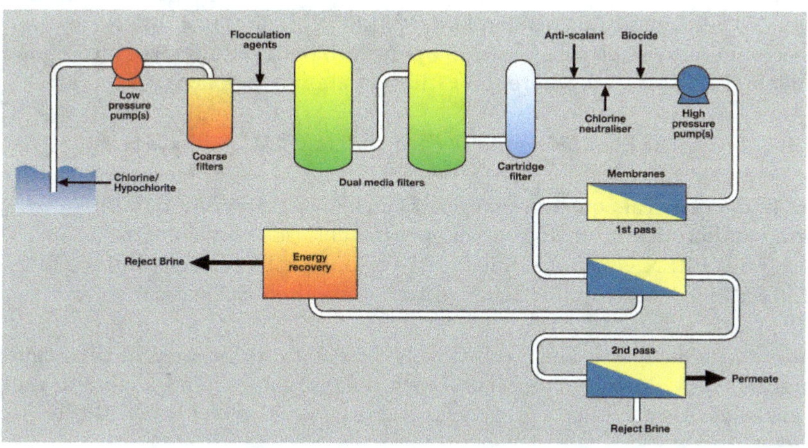

FIGURE 6.1 Schematic diagram of a typical SWRO plant.

6.1 Pretreatment Section

The aim of this process part is to generate the best seawater quality for sustainable membranes by eliminating especially suspended solids and biofouling forming species (organic molecules, microorganisms). Except filtration, which is a physical process, the remaining typical pretreatment operations are mostly chemically based and include chlorination, coagulation, acidification or antiscaling, and dechlorination. They aim to reduce the biofouling and scaling, and to preserve the RO membranes from chemical damage.

From a corrosion perspective, the salinity of the feed seawater ranges globally from 3.5–4% in the Arabian Gulf, and the annual seawater temperature varies globally from 9–40°C. The interrelation between salinity, temperature, oxygen, and biofouling as it affects corrosion process kinetics on metallic components, is well established and is discussed in this section.

On the other hand, the chlorination (hypochlorite), coagulation ($FeCl_3$ or organic polyelectrolyte), acidification (H_2SO_4), or antiscalant (organic acids based) injection, and dechlorination (sodium bisulfite) generate a complex corrosive condition with a redox potential, which ranges from 250–350 mV saturated silver/silver chloride ($Ag/AgCl_{SAT}$, where SAT indicates saturation). The effect of this redox potential is not limited to the RO membrane degradation but also the stainless steel electrochemical potentials, as discussed later.

In terms of metals, copper alloys are not often used in this section because of their negative effect on membrane performance and produced water quality. However, nickel aluminum bronze (NAB) has been used for seawater lift pumps in some plants.

Much more common is the use of stainless steels, and especially those that are resistant to pitting and crevice corrosion in seawater. In this regard, standards such as the Norwegian oil and gas standard (NORSOK),[1] states that for aerated seawater service, stainless steels and nickel alloys must have a pitting resistance equivalent number (PREN) > 40.0, where the PREN equation is as follows:[2]

$$\text{PREN} = \%\text{Cr} + 3.3 \times (\%\text{Mo} + 0.5 \times \%\text{W}) + 16 \times \%\text{N}$$

This is an empirical equation that relates an alloy's composition to its resistance to pitting and crevice corrosion in near-neutral chloride-containing solutions. There are several PREN equations, but this is the one that has been adopted by NACE MR0175/ISO 15156[(1)] and is being written into other ISO standards.

If lower grades of stainless steels (PREN <40.0) are to be used in this section of the plant, they should be associated with reliable cathodic protection to mitigate localized corrosion risks.

1. See the Bibliography at the end of this chapter for full reference information.

Coated carbon steel may be used for some applications, but the quality of the coating is essential to guarantee a reasonable service life. Such coatings include synthetic rubbers like neoprene, or an epoxy-based coating such as glass flake epoxy.

High chromium and molybdenum-nickel alloys, such as alloys 625 and C-276, and titanium and its alloys, are rarely used because of their high cost. Fiberglass reinforced plastic (FRP) is used in larger plants for low pressure seawater piping, particularly in larger diameters and where there are long runs with few bends and tees.

The main components will be discussed in terms of corrosion conditions and materials selection.

6.1.1 Pumps

In older plants, NAB has occasionally been used for seawater pumps, but accelerated corrosion can occur with NAB if it spends some time in stagnant seawater and sulfate reducing bacteria (SRB) become active. If NAB pumps are used, it is imperative that they are not connected directly to high alloy stainless steel piping, as this will cause accelerated galvanic corrosion of the NAB because of its lower corrosion potential,[3] as discussed in the galvanic corrosion section in Chapter 3.

It is well established that cast and wrought stainless steels such as 2205 duplex and 25%Cr duplex stainless steels with PRENs from 34–37, can suffer crevice corrosion in seawater in SWRO plants.[4] Hence, for lift pumps, superduplex stainless steel should be cast to ASTM A995, grade 5A or 6A. Because of its composition, grade 5A tends to be more ferritic and therefore more prone to cracking in thicker sections. Francis and Hebdon discuss the merits of castings in both grades and conclude that larger castings are preferred in grade 6A.[5] However, the *quality* of the castings is the most important factor, and this is discussed in more detail in Chapter 11. Francis and Hebdon also discuss castings in the 6%Mo austenitic grades and point out the problems with obtaining sound castings.[5] In addition, the lower strength and higher molybdenum and nickel contents of these alloys make castings more expensive than those in one of the superduplex grades.

In larger SWRO plants, the seawater lift pumps were sometimes made of austenitic cast iron with stainless steel impellers and shafts. However, there have been numerous failures of such pumps by stress corrosion cracking (SCC) especially in warmer waters, as described in Chapters 4 and 5. Many of these have been replaced with superduplex stainless steel pumps, and the use of austenitic cast iron for seawater pumps has declined.

Severe corrosion of the low pressure superduplex pump casing and shaft was observed in an SWRO plant (Figure 6.2). The causes of the failure were mainly due to poor quality related to the composition and microstructure. The need to control these to obtain good quality duplex and superduplex castings is discussed in more detail in Chapter 11.

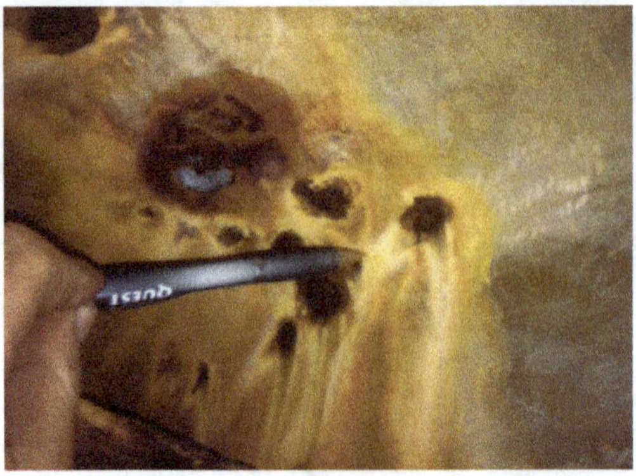

(A)

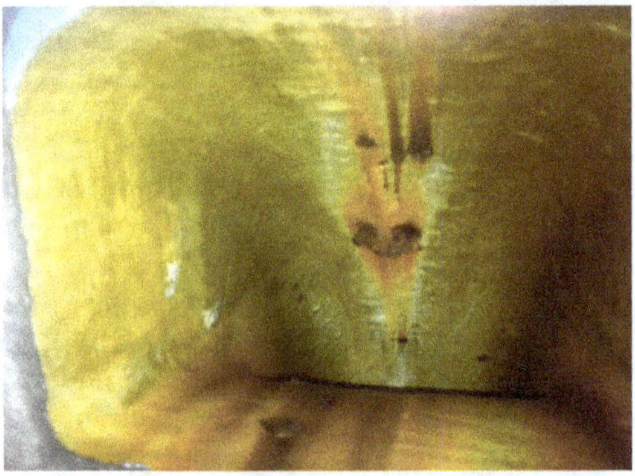

(B)

FIGURE 6.2 Severe corrosion of superduplex stainless steel castings due to poor quality control: (A) pitting at major weld repair, and B) pitting at sigma/chi phase due to poor heat treatment.

6.1.2 Valves

Typical valves in this section are usually butterfly, for large diameter piping in concrete, or FRP that are rubber-lined steel with a NAB disk. Smaller valves, as well as those fitted in high alloy stainless steel piping, should be in a high alloy stainless steel to avoid galvanic corrosion problems, and superduplex is usually the most cost-effective choice. One thing that is sometimes neglected is minor valves, such as those used in drains. They are often positioned at the bottom, with a short pipe between the main system and the valve. The solution to avoid galvanic corrosion failures is to use high alloy stainless steel (PREN >40) for all minor valves in contact with seawater.

Stagnant conditions are strictly prohibited for NAB valves because they often induce high sulfides produced by SRB activity. For instance, it is recommended to avoid having the valve spend excessive times in the closed position while full of seawater. Larche et al. report the failure of a NAB disk in a coated steel drain valve due to sulfide pitting.[4] Moreover, galvanic coupling between NAB and high alloy stainless steel should also be avoided.[3]

The importance of selecting galvanically compatible valves was stressed above, and Figure 6.3 shows severe galvanic corrosion of an NAB drain valve fitted in a superduplex stainless steel high pressure line.

FIGURE 6.3 Galvanic corrosion of an NAB drain valve in a superduplex stainless steel high pressure line. *Photo courtesy of Rolled Alloys*

6.1.3 Strainers and Filters

Strainers and filters can be subjected to high local stresses when there are excessive entrained solid bodies in the feed seawater. For this reason, copper alloys are not widely used for inlet strainers. If 316L stainless steel is used, note that it will suffer crevice corrosion in seawater, so it must be protected by cathodic protection. However, it is difficult to make a large cathodically protected filter system that is robust enough to survive without regular maintenance. The alternative is to use superduplex stainless steel, which is readily available as wires and mesh. In addition, thin sheet can be drilled or punched to make a filter with a specific pass size. All of the support structure can be superduplex or polymers, where a lower strength is acceptable.

The dual media filter vessels can be large (several meters in diameter and height) and it is common to fabricate them from carbon steel and coat all internal surfaces with a synthetic rubber, such as neoprene, offering a finite life (10–15 years). To avoid relining, crevice corrosion resistant stainless steels (superduplex or 6%Mo austenitic) can be a cost-effective alternative, either as a solid or clad onto steel.

The final filters (typically 1 μm) are usually polymeric; the bodies may be polymer as well, or carbon steel with a good quality coating, such as neoprene. For smaller filters, coating carbon steel is difficult, and in such cases superduplex or 6%Mo austenitic stainless steel are suggested.

One typical example of final filter corrosion failure is shown in Figure 6.4. The body was made of stainless steel with PREN below 40 (904L), which suffered from crevice corrosion.[6] Another example is the short time (six weeks of operation) failure of large, dual-media filter vessels made of duplex 2205 (PREN = 35). They began to leak because of pitting at welds (Figure 6.5); crevice corrosion was also observed at flanged joints.

FIGURE 6.4 Crevice corrosion of 904L final filter top-flange basket.

FIGURE 6.5 Leaks due to pitting at weld of 2205 duplex dual media filter vessel.

In another SWRO plant in the Middle East, some low pressure superduplex stainless steel pumps showed crevice corrosion after entering service, with a seawater temperature of 25–30°C (Figure 6.6). Unfortunately, the pumps were subject to stagnant seawater for the 6–9 months of final construction and commissioning, and were only run intermittently with very infrequent chlorine dosing. In addition, the pumps were left full of seawater when idle and SRB became active. The full investigation can be read in Larche et al.[4]

FIGURE 6.6 Crevice corrosion of superduplex low pressure pump case after extended stagnation prior to service.

6.1.4 Pretreatment Chemical Systems

Chemicals used for pretreatment are stored in tanks made of rubber lined carbon steel, FRP composite, or stainless steel. In the authors' experience, the advice given from chemical suppliers is not always reliable and has resulted in field failures. For instance, despite 316L being recommended for storing an acidic biocide (pH~2.3), attack occurred at the waterline. Corrosion tests showed that 316L would resist corrosion in the bulk chemical but was susceptible to attack at the waterline at both 5% and 20% concentration, at 30°C (typical water temperature in the Middle East). However, 2205 duplex was resistant to waterline attack at 5% but not at 20%, and superduplex Z100 was resistant at both concentrations. If FRP is used for storage tanks, it is essential that a resin and veil are chosen that will resist the specific chemical being stored, e.g., glass is unsuitable with alkaline chemicals.

Transport of pretreatment chemicals to the dosing point is usually done with pipes made of austenitic 316L stainless steel or plastic, such as chlorinated polyvinyl chloride (CPVC) or unplasticized PVC (uPVC). The same materials are adopted for the isolation valves immediately prior to the injection point; superduplex stainless steel instead of 316L should also be considered because part of the valve will be in contact with seawater when closed.

6.2 High Pressure Section

All the filtration and treatment of the water for the high pressure section makes it less corrosive than natural seawater. The membranes will degrade if the redox potential exceeds 350 mV Ag/$AgCl_{SAT}$. One way to achieve very low redox potentials is to dose with sodium metabisulfite, however, this method is expensive so the practical lower limit is 250 mV Ag/$AgCl_{SAT}$. In most SWRO plants, the redox potential is close to the upper limit. Byrne et al. showed that the range of 250–350 mV Ag/$AgCl_{SAT}$ corresponds to an open circuit potential for high alloyed stainless steels of 100–200 mV SCE[(2)] in synthetic seawater.[7] In practice, stainless steel in most plants will have a potential close to the upper limit.

Figure 6.7 shows the wide range of potentials that are possible for stainless steels in seawater.[7] In natural seawater, a biofilm forms over a period of a few days to a few weeks causing the ennoblement of the potential up to +300 mV SCE. Chlorine additions (~0.5 mg/L) prevent the formation of a biofilm, but chlorine is a powerful oxidizer and the potential rises to around +600 mV SCE. When seawater is heated 25–30°C above its normal ambient, the biofilm does not form, and the potential is around +150 mV SCE. Biofilm development temperature limit depends on the seawater world location, as discussed by Meroufel.[8] Conversely, deaeration (partial or full oxygen removal) shifts the potential in the negative direction. The red lines in Figure 6.7 show the maximum and minimum potentials for stainless steels in the high pressure section of an SWRO plant. The consequences of these electrochemical data from a crevice corrosion perspective were studied by Byrne et al.,[7] who determined the critical crevice temperatures for some stainless steels over a range of constant potentials.

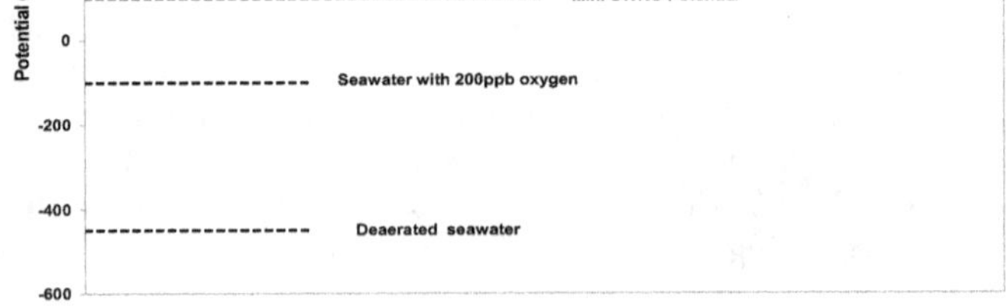

FIGURE 6.7 Possible potentials for stainless steels in seawater.[7]

2. SCE = saturated calomel electrode

The results from Bryne et al.[7] (Figure 6.8), show that alloy 316L is totally unsuitable in seawater at any potential down to -100 mV SCE. Alloys 904L and 2205 were only resistant up to around 22°C at potentials of 200 mV SCE or greater, but there was a sharp increase in the critical crevice temperature (CCT) at lower potentials. Most applications of 904L and 2205 in the high pressure section of SWRO plants have failed because of crevice corrosion and/or pitting of welds,[7, 9] because most plants operate at higher temperatures than 22°C. Where these alloys have worked, the temperatures were lower, or the redox potential was closer to the lower limit shown in Figure 6.7. Therefore, despite all the pretreatment of the seawater, only superduplex and 6%Mo austenitic alloys will perform satisfactorily under the conditions commonly found in the high pressure section.

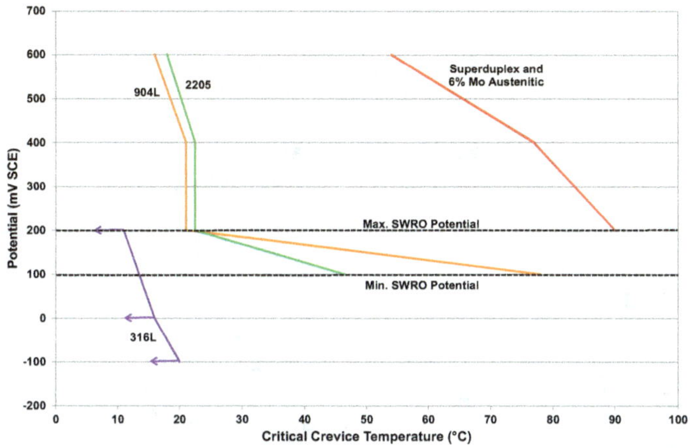

FIGURE 6.8 Critical crevice temperatures for some stainless steels as a function of potential.[7]

Byrne et al. also investigated the effect of chloride on the CCT at different potentials for 2205 and Z100 superduplex stainless steel, as shown in Figure 6.9.[7] It shows that the CCT is hardly affected at all by chloride, with a reduction of only 10–20°C over the range 3,000–100,000 mg/L chloride. It can also be seen that lower potentials produce a much greater increase in CCT for superduplex compared with 2205. It is expected that the results for the 6%Mo alloys should be similar to superduplex, while those for 904L should be similar to 2205.

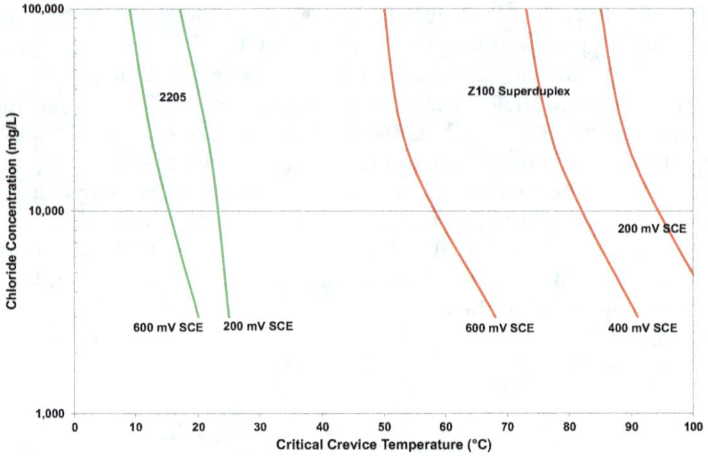

FIGURE 6.9 Critical crevice temperatures (CCTs) for some stainless steels as a function of chloride.[7]

6.2.1 Pumps

Pumps in the high pressure section are multistage to deliver the high pressure required (60–100 bar [6–10 MPa]) and are either a horizontally split case or a radially split case design for ease of maintenance. These are usually made of superduplex stainless steel because of its high strength, ease of manufacture, and lower cost compared with 6%Mo austenitic alloys. This alloy selection also avoids problems due to crevice corrosion, as discussed previously. As for the low pressure pumps, it is important that the castings are of good quality and the guidelines of Francis and Hebdon should be followed[5] (see also Chapter 11).

6.2.2 Piping

High pressure piping must contain high pressure (up to 100 bar [10 MPa]) and have crevice corrosion resistances. After initial bad experiences with 316L and 317L, the SWRO industry moved directly to the 6%Mo austenitic stainless steels. Then, with the development of the superduplex grades, they became cost competitive and were adopted widely. Because of the need to be able to remove any membrane cell for maintenance or replacement, pipe connections are made with high pressure couplings, as shown in Figure 6.10. Each pipe stub has a square cut groove about 15 mm from the end, and a synthetic rubber boot joins the two pipes. Two half clamps fit in the grooves and are bolted together to make a seal. These clamps are usually galvanized steel or 316L stainless steel.

FIGURE 6.10 A typical high pressure coupling with one half-clamp removed for clarity.

The boot creates a tight crevice and lower alloy stainless steels (PREN <40) have suffered severe crevice corrosion in SWRO plants, as shown in Figure 6.11. Superduplex stainless steel would be expected to resist crevice corrosion under the boot of high pressure couplings, but Byrne et al. report corrosion there of Z100 superduplex in one plant.[7] They carried out crevice corrosion tests on the outside of samples of Z100 superduplex pipe with different surface finishes at +200 mV SCE, the maximum stainless steel potential in the high pressure section. The results in Figure 6.12 show that the maximum CCT was unchanged with surface finish, but the tube with the outer surface machined showed a much smaller scatter of CCT values. This is thought to be because machining removed any ovality and roughness, both of which could create local crevices that were much tighter. Machining the land under the boot prevented corrosion there in subsequent SWRO applications.[7]

FIGURE 6.11 Crevice corrosion of 316L under the boot of a high pressure coupling.

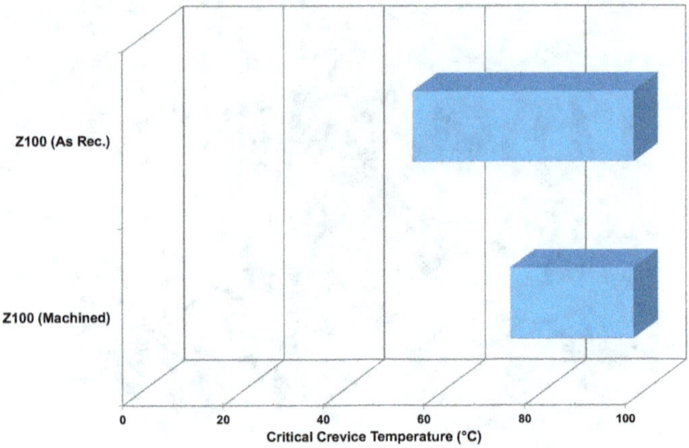

FIGURE 6.12 The effect of surface finish on the CCT range of superduplex in seawater at a potential of +200 mV SCE.[7]

Piping in the second pass may be in a lower alloy stainless steel if the chloride concentration is low enough (≤4,000 mg/L), and 2205 has been used successfully in larger plants. However, the amount of second pass piping in smaller plants is small and it is often more convenient to make it all in superduplex stainless steel, especially because it removes the risk of the wrong grade being used in the first pass.

In SWRO piping, it is frequently required to take a small branch from a large diameter pipe, for example, NPS1.5 from an NPS8 main pipe. One way is to drill a small hole in the large pipe and pull a small branch to weld the smaller pipe to. Superduplex stainless steel is difficult to pull in schedule 40S pipe and tearing may occur. An alternative is to drill a hole in the large pipe and fillet weld the branch to it. This is difficult to do, even in the fabrication shop, and corrosion has been seen in such fillet welds. Such a procedure really needs a corrosion test as part of the qualification to ensure the joint will be sound, as discussed in Chapter 11. The preferred method is to use a reduced branch tee fitting, which will only require regular butt welds to fit into the piping, but this is rarely done in practice because of the added cost.

It is also common to use threaded connections for things such as drain plugs and sensors. Such joints create tight crevices and have been known to suffer crevice corrosion. It is important to use a screwed connection in superduplex stainless steel and seal the threads with both PTFE(3) and a suitable sealant, such as Loctite,(4) to ensure water does not enter the threads.

3. PTFE = polytetrafluoroethylene
4. Loctite is a registered trademark of the Henkel Corporation.

In SWRO plants, spiral wound gaskets are sometimes used, especially in the high pressure section. The metal for the windings must be compatible with the flange alloy, which usually means, superduplex stainless steel, 6%Mo stainless steel, Ni-Cr-Mo alloys with Mo > 7%, or titanium.[10]

Sometimes a 316L spiral wound gasket has been used but coated with PTFE to prevent corrosion of the windings. PTFE forms a much tighter crevice than a bare metal spiral wound gasket, and it can cause crevice corrosion when it would not be expected, as shown in Figure 6.13.[10] The conclusion is that PTFE coated gaskets should be avoided.[10]

FIGURE 6.13 **Crevice corrosion of superduplex stainless steel flange under a PTFE coated spiral wound gasket.** ***Photo courtesy of Rolled Alloys***

An unusual problem occurred in the high pressure section of an SWRO containerized plant in the Middle East, where the high pressure couplings were fusion bond epoxy (FBE)-coated carbon steel. Corrosion was seen on the end of the superduplex feed pipes to the membrane cells and on the exit pipes from the cells.[11] This was stray current corrosion, which was thought to be due to the plant's location on land with a high conductivity because it was recovered from the sea, and all the local electrical systems were earthed into this ground. It was demonstrated that the FBE coating acted as an insulator, which meant that much of the metal piping was floating with respect to earth. The problem was solved by anchoring all the piping to a common earth.[11] This demonstrates that galvanized steel or stainless steel clamps are the best choice for the high pressure couplings, because these will ensure a continuous electrical connection between all the piping. Stray current was also reported to initiate and accelerate the propagation of crevice corrosion on 316L.[12]

6.2.3 Membrane Pressure Vessels

The membrane pressure vessels are constructed in FRP and are suitably rated for operation up to 83 bar (8.3 MPa). The vessels are designed and manufactured as per ASME Section X. The high pressure feed and reject ports are most often in superduplex stainless steel. The industry standard is to use side-ported vessels with multiple ports. This allows economy in piping. The ports are grooved to allow use of flexible couplings for connection with high pressure piping. The industry is moving toward the use of duplex and superduplex stainless steel couplings with U.S. Food and Drug Administration (FDA)-approved rubber gaskets.

6.3 Brine Rejection Section

The chloride in the reject brine section may be double that in the seawater feed, but this will not affect the choice of materials because superduplex and 6%Mo austenitic stainless steels will be more than adequate, while lower alloy stainless steels will not. This assumes that there have been no changes in the water treatment prior to the high pressure pumps that have significantly changed the redox potential.

The reject brine section is at high pressure and has a high chloride content, but Figure 6.9 shows that this will not cause a significant reduction in the CCT for superduplex and 6%Mo austenitic stainless steels. These alloys are generally used for the piping, but the hydraulic energy recovery devices (ERDs) may be made with ceramic internals, polymers, or superduplex stainless steel, depending on the design. The materials must resist corrosion and erosion, and maintain efficiency. Figure 6.14 shows a typical layout of superduplex reject brine piping coupled to hydraulic ERDs. Couplings in this section are also of the high pressure type shown in Figure 6.10.

FIGURE 6.14 Superduplex brine reject piping coupled to hydraulic energy recovery devices. *Photo courtesy of Rolled Alloys*

6.4 Future Trends

In current SWRO plants, the final filters in the low pressure section generally go down to ~1 μm, but there are advantages in finer filtration, if it can be made reliable and economic. Microfiltration would be down to ~0.1 μm, while ultrafiltration would be to ~0.01 μm and nanofiltration would be to ~0.001 μm. While micro- or ultrafiltration could extend membrane life, they do not remove the need to inject antiscaling chemicals and biocides. The advantage of nanofiltration would be that chemical dosing of antiscalants and biocides prior to the high pressure pumps would be unnecessary and the finer filtration should also prolong membrane life. It is not known if this very fine filtration would further reduce the redox potential to enable lower alloyed stainless steels to be used. The authors think it unlikely that the redox potential would be sufficiently reduced, and crevice corrosion testing would be advisable before going ahead with a low alloy in the high pressure section, in a plant with nanofiltration.

On the other hand, recent efforts toward zero liquid discharge (ZLD) to manage the seawater brine to extract salts and chemicals induce new processes with highly concentrated brine (total dissolved solids up to 100,000 mg/L) at temperatures typical of the Arabian Gulf region (30–40°C). While these processes remain at pilot scale, corrosion data on materials that can resist those aggressive conditions are missing. Indeed, highly concentrated brines should be almost free of oxygen; a low concentration of oxygen is necessary for the stability of the passive layer of low alloy grades of stainless steel.

References

1. NORSOK M-001 (latest edition), "Materials Selection," Edition 5 (Oslo, Norway: Standards Norway, September 2014).
2. Olsson, C-O.A., "Wet Corrosion of Stainless Steels and Other Chromium Bearing Alloys." In: *Encyclopedia of Interfacial Chemistry*, K. Wandelt, ed. (Amsterdam, Netherlands: Elsevier, 2018): p. 535.
3. Francis, R., Galvanic Corrosion: *A Practical Guide for Engineers*, 2nd ed. (Houston, TX: NACE International, 2017).
4. Larche, N., P. Dezerville, and D. Le Flour, "Corrosion and Corrosion Management Investigations in Seawater Reverse Osmosis Desalination Plants," *Desalination and Water Treatment* 51 (2013): pp. 1744–1761.
5. Francis R., and S. Hebdon, "The Selection of Stainless Steels for Seawater Pumps," Corrosion 2015, paper no. 5446 (Houston, TX: NACE International, 2015).
6. Meroufel, A., "Crevice Corrosion within RO Desalination Plants—Case Studies and Recommendations," Jubail Corrosion and Materials Engineering Forum (Jubail, Saudi Arabia: NACE Jubail Saudi Arabia Section, 2017).
7. Byrne, G., R. Francis, G. Warburton, and J. Wilson, "Electrochemical Potential and the Corrosion Resistance of Stainless Steels in SWRO Applications,"

IDA International Desalination Conference (Topsfield, MA: International Desalination Association, 2009).
8. Meroufel, A., "Corrosion in Reverse Osmosis Desalination Processes: Forms and Mitigation Practices." In: *Corrosion and Fouling Control in Desalination Industry*, V.S. Saji, A.A. Meroufel, and A.A. Sorour, eds. (Cham, Switzerland AG: Springer Nature, 2020).
9. Olsson, J.O., and M.M. Snis, "Don't Repeat Mistakes! An SWRO Plant Case Study," IDA International Desalination World Congress, paper no. SP050-036 (Topsfield, MA: International Desalination Association, 2005).
10. Francis, R., and G. Byrne, "Gasket Selection for Stainless Steels in Seawater," *Materials Performance* 46, 10 (2007): pp. 50–55.
11. Byrne, G., R. Francis, and A. Gulamhusein, "Stray Current Corrosion in Sea Water Reverse Osmosis Systems," Proceedings of the International Desalination Association World Congress (Topsfield, MA: International Desalination Association, 2015).
12. Lenard, D.R., and J.G. Moores, "Initiation of Crevice Corrosion by Stray Current on Stainless Steel Propeller Shafts," *Corrosion* 49 (1993): pp. 769–775.

Bibliography

1. NACE MR0175/ISO 15156 (latest edition), "Petroleum and Natural Gas Industries, Materials for Use in H_2S-Containing Environments in Oil and Gas Production, Part 3: Corrosion Resistant Alloys" (Houston, TX: NACE).

7 Corrosion Under Nonoperating Conditions

Corrosion engineers are aware about the corrosion journey, which starts practically prior to the plant production phase. From the design, construction, through to the operation phase, corrosion engineering should follow all activities. In addition to the corrosion factors related to the QA/QC of materials and equipment (Chapter 11), the aim of this chapter is to outline corrosion that initiates during commissioning or shutdown periods to avoid undesirable failures in the early life of the desalination plant.

Usually, commissioning is focused on the process and how to confirm that the desalination plant can produce the desired drinking water quality in a reliable manner. From the precommissioning to the plant performance test, a lot of activities create corrosion risk situations that should be well understood and properly conducted.

7.1 Precommissioning

During this phase, equipment installation is checked based on specifications where the electrolyte (usually water) is not in continuous contact with material. Dimensional, nondestructive, hydraulic, functional, and performance inspection tests should be carried out in the presence of a qualified team, who should be well aware of related corrosion risks. Then, the critical issues that must be considered from the equipment failure point of view include mainly pipes, pumps, and instrumentation checking. For instance, pipe joint alignment (mechanical, welds) and pump vibration represent the most critical and costly issues that lead to environmentally assisted cracking failures if not fixed properly. ASME B31.3 defines alignment limits of the pipe joints with some tolerance of misalignment, which must be checked from the stress corrosion cracking (SCC) or corrosion fatigue risk point of view.[1] This can be more critical with lower grades of stainless steel in contact with a saline medium. Misalignment in pumps is another chronic issue, causing about 50% of pump failures affecting mainly bearings, shafts, and mechanical seals.

Also, all equipment (pressure vessels, pipes, valves, heat exchangers, etc.) that are subjected to internal pressure or vacuum during operation shall be subjected to hydrostatic testing, as stipulated in the ASME Boiler and Pressure Vessel Code, for example. Depending on the water used for hydrostatic testing and the material in contact, corrosion can initiate during this stage. A careful procedure should be followed that considers the need for chemicals such as biocide, oxygen scavenger, or corrosion inhibitor. Draining is a critical step after completion of the test, where any remaining stagnant water can be detrimental. While ASTM E1003 describes the hydrostatic testing procedure, the previous corrosion concerns are not discussed.[2] However, API 570 recommends some practices for 300 series stainless steel when hydrotested.[3] For instance, the recommended water is a potable one and if it is not available, a high pH (> 10) should be considered together with a suitable corrosion inhibitor. Chloride limits in the test water vary from 50–200 ppm (for lower stainless steel grades) up to 4,000 ppm if duplex 2205 is used. Temperature is another factor that must be considered together with other corrosion factors. NORSOK L-004 also provides some recommendations regarding flushing, cleaning, and pressure testing.[4] Corrosion failures caused by bad practices of hydrotesting are not limited to desalination, and they have been observed in other industrial sectors, for example oil and gas (O&G), and power plants.[5-8] Depending on the material, different corrosion forms were observed due to poor hydrotesting procedure, including pitting, crevice corrosion, microbiologically influenced corrosion (MIC) and SCC.

7.2 Commissioning

This section discusses wet commissioning (water introduction into the systems), process commissioning (chemicals introduction), and performance tests (which can last up to 30 days). The critical chemicals to control during commissioning are biocides and acids, and wherever instrumentation and monitoring are in use, they must be highly accurate. Improper control of these added chemicals can induce severe localized corrosion, especially on corrosion resistant alloys (CRAs).

With large plants it is not uncommon to commission them in stages as specific parts are completed. It is important that actions to prevent corrosion between the commissioning stages are taken, because these intervals may be quite long (e.g., months). A large seawater reverse osmosis (SWRO) desalination plant in the Middle East was commissioned in stages with intervals of a month or more between the stages. The pumps were left full of seawater throughout the commissioning. The low pressure seawater pumps were made of superduplex stainless steel, and the seawater was about 30°C and biologically active. Dosing with biocide was very sporadic, and at full startup there was severe crevice corrosion in the pumps leading to leakage (Figure 7.1). The failures were ascribed to the action of sulfate reducing bacteria (SRB) during the static periods, preventing the formation of the normally protective passive film on startup.[9] This was demonstrated in laboratory tests, where superduplex exposed in seawater with active SRB for 30 days at 30°C suffered subsequent crevice corrosion when exposed to clean flowing seawater.[10] Samples that were preexposed in sterile seawater subsequently resisted crevice corrosion in clean flowing seawater.[10]

FIGURE 7.1 Severe crevice corrosion of a superduplex pump at the split case flange joint after months of being stagnant in biologically active seawater at ~30°C.

The pumps should have been drained and flushed with fresh water as described in the previous section. Similarly, evaporators, flash chambers, and heat exchangers should be drained and flushed as described in Section 7.1 when there are extended intervals between commissioning stages. Alternatively, they can be flushed with fresh water and then left full of fresh water with a suitable biocide and a corrosion inhibitor.

7.3 Shutdown and Layup

Usually, and with the growing water demand, it is rare to observe a desalination plant in shutdown condition for a long time. The annual or periodic shutdown for preventive maintenance does not exceed one to two months, and it is planned in winter when the water demand is less. However, even if the plant is only shut down for a short time, attention should be paid to critical parts to avoid unexpected troubles when restarted. Typical problems after shutdown include sticking of rotating parts, severe corrosion, clogging in small pipes, etc. Operations and maintenance (O&M) manuals describe three types of shutdowns, i.e., short-term (1–10 days), middle-term (10–30 days), and long-term (over 30 days). For the long-term shutdown, a complete drainage, flushing with fresh (or distilled) water, and followed by drying are the main activities. For the middle-term shutdown, only freshwater cleaning is conducted. In the event of a layup, for example when a plant is not scheduled for operation within six months or more, dry store by dehumidifier should be considered for multistage flash (MSF) and multiple effect distillation (MED) evaporators to prevent water condensation and subsequent corrosion.

It may happen that more attention is paid during shutdown to the metallic parts, and nonmetallic components, such as fiberglass reinforced plastic (FRP) pipes are neglected. However, practical examples have shown that leaving an FRP pipe with some seawater in it for weeks can induce a massive growth of SRB bacteria, producing significant quantities of sulfide. A large MSF plant suffered a series of extended shutdowns for maintenance and problem fixing in the steam generation plant. Some of these were up to six weeks. On startup, leakage due to sulfide pitting was soon seen in the copper-nickel heat rejection tubes (Figure 7.2). The reason for this was that the large diameter (> 1 m) FRP pipes feeding seawater to the heat rejection section were several hundred meters long and SRB became active during the shutdown, producing up to 15 mg/L sulfide. The corrosion started when the sulfide-polluted water was flushed through the system at startup, but sulfide continued to enter the flowing seawater as sulfide was slowly leached from the various dead legs in the FRP pipe. Although this produced low levels of sulfide in the water, published data show that as little as 0.03 mg/L sulfide would reduce the corrosion resistance of Cu-Ni tubes in seawater.[11]

FIGURE 7.2 Sulfide pitting of copper-nickel tubes from heat rejection section of an MSF plant.

Another failure case study was recorded in an SWRO plant in brine reject piping made of stainless steel 317L, which was laid up for three years. Preservation of the piping was conducted before plant commissioning, where the pipe was filled with seawater to which formalin (aqueous formaldehyde) was added. Formalin addition was recommended by the project contractor to preserve membranes and other SWRO system components. Significant pitting developed at weld joints (Figure 7.3). However, it seems that formalin induced acidification causing severe pitting of the 317L.[12]

FIGURE 7.3 Pitting corrosion on the weld joint of 317L brine rejection pipe in an SWRO plant due to poor layup procedure.

The over-arching conclusion is that at any stage where components are in contact with static fluids for extended periods (from one day to several months) it is essential to consider the corrosion implications for all alloys. It will often be necessary to drain and flush with fresh water, and may also require wet storage with biocides, corrosion inhibitors, etc.

References

1. ASME B31.3 (latest edition), "Process Piping," ASME Code for Pressure Piping, B31 (New York, NY: American Society of Mechanical Engineers, 2017).
2. ASTM E1003 (latest edition), "Standard Practice for Hydrostatic Leak Testing" (West Conshohocken, PA: ASTM International, 2013).
3. API 570 (latest edition), "Piping Inspection Code: In-service Inspection, Rating, Repair, and Alteration of Piping Systems," 4th ed. (Washington, DC: American Petroleum Institute Publishing Services, 2016)
4. NORSOK L-004 (latest edition), "Piping Fabrication, Installation, Flushing and Testing" (Oslo, Norway: Standards Norge, 2010).
5. Kobrin, G., S. Lamb, A.H. Tuthill, R.E. Avery, and K.A. Selby, "Microbiologically Influenced Corrosion of Stainless Steels by Water Used for Cooling and Hydrostatic Testing," Nickel Institute Technical Publication number 10085 (Toronto, Canada: Nickel Institute, 1997).
6. Prasad, R., "Chemical Treatment Options for Hydrotest Water to Control Corrosion and Bacterial Growth," Corrosion 2003, paper no. 03572 (Houston, TX: NACE International, 2003).
7. Holden, J., A. Hansen, A. Furman, R. Kharshan, and E. Austin, "Vapor Corrosion Inhibitors in Hydro-Testing and Long-Term Storage Applications," Corrosion 2010, paper no. 10405 (Houston, TX: NACE International, 2010).

8. Zheng, W., W.R. Tyson, R.W. Revie, G. Shen, and J.E.M. Braid, "Effects of Hydrostatic Testing on the Growth of Stress-Corrosion Cracks," International Pipeline Conference, Vol. 1, paper no. 2053 (Calgary, Canada: ASME Pipeline Division, 1998).
9. Larche, N., P. Dezerville, and D. Le Flour, "Corrosion and Corrosion Management Investigations in Seawater Reverse Osmosis Desalination Plants," *Desalination and Water Treatment* 51, February (2013): p. 1744.
10. Ohashi, K., R. Kobayashi, J.F.D. Stott, and M.J. Schofield, "Marine Crevice Corrosion of Stainless Steels Under Biofilmed and Sterile Conditions," paper no. 7109, Corrosion 2016 (Houston, TX: NACE International, 2016).
11. Francis, R., "Effect of Pollutants on Corrosion of Copper Alloys in Seawater; Part 2–Sulphide and Chlorine," *British Corrosion Journal* 20, 4 (1985): p. 175.
12. Malik, A.U., I. Andijani, S. Al-Fozan, F. Al-Muaili, and M. Al-Hajri, "An Overview of the Localized Corrosion Problems in Seawater Desalination Plants: Some Recent Case Studies," IDA World Congress, paper no. IDAWC/MP07–162 (Topsfield, MA: International Desalination Association, 2007).

8 Corrosion Mitigation Practices

In this section, all corrosion mitigation activities in desalination plants will be summarized for better visibility for asset managers. Mitigation practices can be categorized from different corrosion perspectives, i.e., design, material selection, and environment control. The first two were developed with accumulated operation and maintenance experience across desalination plants around the world. However, environment control has been already established in other seawater handling industrial processes, such as oil and gas, and power plants.

8.1 Design

Corrosion mitigation starts at the design phase of all industrial equipment. In the case of desalination, this can be observed mainly for heat exchangers and evaporators in thermal desalination, and pumps and valves for both thermal and membrane-based desalination processes. Table 8.1 summarizes the main corrosion risks, and related corrosion threats with design action to mitigate and reduce the corrosion effect. It may happen that a corrosion engineer will choose a less expensive design with reliable materials available in the market. This is the case, for instance, for seawater valves with the dominance of butterfly valves known to be less costly compared to the ball valves that offer better operational performance.[1] Butterfly valves also offer an additional design benefit with the relative ease of internal lining of the valve body, compared to other designs, such as gate or globe.

For pumps, many efforts are undertaken by manufacturers to produce pumps with a design that offers the best life cycle cost through minimum maintenance. This is also the case for energy recovery devices (ERDs) where the pressure exchanger (PX) isobaric model is considered as a fail-safe design, contrary to the piston isobaric ERDs, which requires periodic maintenance.[2]

TABLE 8.1. Corrosion mitigation in desalination plants by design.

DESALINATION PROCESS	EQUIPMENT OR PART	RISK	THREATS	DESIGN ACTION
MSF	Heat recovery tubes	Vapor side corrosion	CO_2, O_2, Temperature	Multiple noncondensable gas venting points
	Flash chamber	Pitting corrosion	Stagnant seawater or brine	Floor slope and draining point locations
MED	Effect tubes	Underdeposit corrosion	Dry zones	Spraying system enhancement
Thermal and RO[a]	Pump	Cavitation	Pump operation	Increase NPSHa[b] or reduce NPSHr[c]

a. RO = Reverse osmosis
b. NPSHa = Net positive suction available
c. NPSHr = Net positive suction required

Another design aspect that plays a role to mitigate corrosion is the operational performance. This is the case, for instance, of heat recovery tubes in multistage flash (MSF) plants in terms of noncondensable corrosive gases (e.g., CO_2, O_2). CO_2 is known to affect vapor-side corrosion, as well as the heat transfer performance. However, most of the venting systems are designed with a cascade configuration between stages, based on the estimated quantity of CO_2 release.[3] Practical experience has shown better reliability and longer life for the units with multiple suction points instead of the cascade configuration widely seen in many desalination units.

Another factor is that the size of heat recovery tubes is so big in many cases (~20 m length) that the high velocity circulating brine induces significant and damaging vibration, requiring the optimization of the spacing of tube supports and baffles.

8.2 Materials Selection

Materials selection in desalination plants is controlled by several criteria depending on the concerned equipment or part. This includes corrosion resistance, thermal conductivity, mechanical strength, welding, and cost. Improper materials selection is known to result in costly failures that can be detrimental to production and cause extra work for operations and maintenance (O&M) staff. Different approaches and tools are proposed for an efficient material selection;[4] however, in the present section, only the criteria of materials selection will be briefly discussed from the process and failure risks perspectives.

8.2.1 Corrosion Resistance

The first and dominant material property is that corrosion resistance may vary from one desalination process to another, as well as within the same process. For instance, seawater aggressivity varies depending on its status in the desalination processes. Indeed, its oxygen content, salinity (chlorides concentration), temperature, chlorine residual, suspended particles content, and velocity directly affect the material choice. The aggressivity of vapors present in thermal desalination processes is controlled mainly by temperature, acidity, oxygen content, solids entrainment, and velocity.

Before continuing with the remaining noncorrosion related criteria, it is important to consider the galvanic corrosion compatibility if a piece of equipment will be constructed of different alloys. In this regard, conventional galvanic series in seawater at different temperatures is an important tool to be considered initially. Then, cathodic current data based on experimental or modeling studies are considered to assess the risk for long-term operation. In some cases, galvanic corrosion cells are allowed, such as for brine recycle pumps where an austenitic cast iron case provides long life for the stainless steel internals. Further information on galvanic corrosion and methods of avoiding it are given in Reference 5.

The additional corrosion risk during typical maintenance activities should also be considered. An example is the case of acid cleaning of heat exchangers when the production performance is affected by the buildup of scale.[6]

8.2.2 Thermal Conductivity

Thermal conductivity is required for heat transfer function, as is the case for heat exchanger tubes present in thermal desalination plants (MSF and multiple effect distillation [MED]). Table 8.2 summarizes typical heat exchanger tube materials with their thermal conductivities. Alloys with lower thermal conductivities, such as grade 2 titanium and stainless steels, are available with thinner walls than the copper alloys to improve overall heat transfer.

TABLE 8.2. Thermal conductivity and mechanical properties of typical heat exchanger tube materials.

ALLOY	THERMAL CONDUCTIVITY (W/mK)	YIELD STRENGTH (MPa×10^3)	MODULUS OF ELASTICITY (GPa)	HARDNESS (HRB)
90/10 Cu-Ni	40	138	124	20
70/30 Cu-Ni	29	159	152	22
Al-brass	100	186	117	83
Ti Gr2*	24.5	310	106	92
S44660**	17.48	480	215	95

* Ti Gr2 = Titanium Grade 2, **Super ferritic stainless steel

8.2.3 Mechanical Strength

Depending on the type of mechanical load (tension, compression, shear, etc.), the total stress level (residual stresses in the alloy, fabrication stresses, and application stresses) must be considered. It is also necessary to consider whether the stress is static or dynamic. If it is dynamic, it will be necessary to consider the resistance to fatigue and corrosion fatigue. The main equipment affected by mechanical strength requirements are pumps (all desalination processes) and pressurized pipes (mainly in reverse osmosis [RO] desalination). Components requiring consideration of fatigue resistance are pump shafts and heat exchanger tubes. For example, titanium tubes require a shorter baffle spacing than copper alloys because of their higher susceptibility to fatigue.

8.2.4 Welding

Welding criteria may be considered from either manufacturing (weldability) or joining perspectives. Welding, in terms of joining, was discussed in Chapter 2 with regards to the corrosion risks. From the alloy manufacturing side, weldability corresponds to the ability of a material to be welded under a particular set of conditions. Examples of materials used in desalination that can be considered as nonweldable are rare, but they do exist, e.g., austenitic cast iron.

8.3 Environment Control

Aggressive environments in desalination plants can have either a liquid (seawater, brine) or vapor form. The control of the environment is not possible for all three because of process operating conditions. Nevertheless, the control of some of their characteristics regarding both corrosion and fouling is well established. For instance, seawater corrosivity starts from its oxygen content, with the effect depending on the material in contact. While a low dissolved oxygen concentration is desirable for carbon steel and other active alloys, this can be detrimental (especially in stagnant conditions) to the stability of the passive layer on the surface of passive alloys.[7] However, even in fully deaerated waters, high alloy stainless steels (superduplex and 6%Mo) can reduce sufficient water to create a passive film.[8] Seawater *deaeration* is performed only in MSF plants, while it is *aerated* in MED (low oxygen) and RO (fully aerated) desalination processes. These latter processes are known for their wide adoption of corrosion resistant alloys (CRAs).

The microbiological activity in the seawater stream is mitigated by conventional biocide (sodium hypochlorite) injection. This is practiced in all desalination processes. While this method is known to reduce sessile bacteria up to high extent, it induces a side effect by increasing the oxidizing power of the seawater. One of the examples of this side effect is the raising of stainless steel potentials toward the pitting potential region.[9] The limits of use of superduplex and 6%Mo stainless steels in terms of temperature and chlorine concentration have been published.[10, 11]

The injection of antiscalant chemicals is usually discussed only from scale impact on heat transfer (thermal desalination) or membrane separation performance (RO desalination). However, scaling on metallic surfaces is also a corrosion threat. Particularly, the scaling development on heat exchanger tubes induces two corrosion issues: 1) creation of hot spots, which accelerate vapor side corrosion, and 2) the scale can promote underdeposit corrosion.

The seawater or brine in thermal desalination plants has the dissolved salts concentration controlled to prevent the deposition of hard scale (calcium sulfate [$CaSO_4$]). The formation of this scale causes a severe decline in heat transfer, which sharply increases the steam quantity (energy consumption) to maintain distilled water production.

The aggressivity of vapors in thermal desalination plants is controlled through minimizing noncondensable gases dissolution. Air removal and CO_2 extraction are the main actions undertaken in thermal desalination processes because high levels of either one will increase vapor side corrosion.[12]

References

1. Avery, R.E., "The Application of Copper-Nickel Alloys in Marine Systems," CDA Inc. Seminar, Technical Report 7044-1919 (New York, NY: Copper Development Association, 1996).
2. Stover, R.L., "Seawater Reverse Osmosis with Isobaric Energy Recovery Devices," *Desalination* 203 (2007): pp. 168–175.
3. Glade, H., and K. Genthner, "Measurements of CO_2 Release Rates in Individual Stages of MSF Distillers, IDA World Congress San Diego (Topsfield, MA: International Desalination Association, 1999).
4. Rahim, A.A., S.N. Musa, S. Ramesh, and M.K. Lim, "A Systematic Review on Material Selection Methods, *Journal of Materials: Design and Applications* 234 (2020): pp. 1032–1059.
5. Francis, R., *Galvanic Corrosion: A Practical Guide for Engineers*, 2nd ed. (Houston, TX: NACE International, 2017).
6. Meroufel, A., "Corrosion Control During Acid Cleaning of Heat Exchangers." In: *Corrosion and Fouling Control in Desalination Industry* (New York, NY: Springer, 2020): p. 209.
7. Malik, A.U., and S. Al-Fozan, "Corrosion and Materials Selection in MSF Desalination Plants," *Corrosion Review* 29 (2011): pp. 153–175.
8. Newman, R., University of Toronto, Canada, private communication (2005).
9. Olsson, C-OA., "Wet Corrosion of Stainless Steels and Other Chromium Bearing Alloys." In: *Encyclopedia of Interfacial Chemistry* (Amsterdam, Netherlands: Elsevier, 2018): p. 535.
10. Francis, R., and G. Byrne, "Experiences with Superduplex Stainless Steel in Seawater," *Stainless Steel World* (Zutphen, Netherlands: KCI Publishing, June 2004): p. 53.

11. Francis, R., *Avoiding Corrosion in Desalination Plants* (Houston, TX: NACE International, 2019).
12. Oldfield, J.W., and B. Todd, "Vapour Side Corrosion in MSF Plants," *Desalination* 66 (1987): p. 171.

9 Lifetime of Equipment

In this chapter, technical lifetime estimations are proposed when applicable for different equipment in each of the two main desalination processes (thermal and membrane based). It is important to understand how lifetime is predicted because it depends on the material, its quality, the environment, and the corrosion types discussed in detail in previous chapters of this book. To avoid repetition due to some items being common to all types of desalination plants, the information is organized by considering the three main environments and their corrosiveness. The tables are subdivided so that data for one type of plant are gathered in the same place. Where the equipment materials are identical in different plants, some cross referencing has been unavoidable. Equipment or parts in contact with distilled or potable water are excluded because of the low aggressivity of these media.

The alloys are mostly referred to by their generic names (as listed in the appendix). Only the major equipment items are discussed here; the reader should refer to the relevant chapter for details on more minor items.

Throughout the history of the desalination industry, plant designers have adopted what is called the "economic life of a desalination plant," which refers to the useful life when the asset is profitable for the owner. The economic life is estimated by assuming normal operation with preventive maintenance. This economic life was about 20 years in the late 1970s and early 1980s and jumped to 40 years after the accumulation of operation and maintenance experience data.[1, 2] Within this context, corrosion engineers focus more on the physical life of the plant, which is related to the corrosion type and/or rate, and mechanical integrity (allowable stress). It is worthwhile to mention that the physical life includes the economic life and the life in which the asset is not profitable.

The focus of this chapter is on the corrosion part of the equipment's physical lifetime. There is no doubt that field corrosion data are the best to rely on for an accurate estimation. However, this is not always available in the open literature–it can be plant specific, and highly affected by operation and maintenance (O&M) practices. Thus, where a range is quoted, this is because the lifespan depends on how well the item is maintained during its lifetime, and all the alloys are assumed to be supplied in the correct condition. Footnotes have been used with each table where there are important provisos connected with a particular alloy choice.

When lifetime estimation is applicable, plant designers consider a corrosion allowance, which is affected by the corrosion rate and the target design life. The corrosion rate can be either conservative or underestimated depending on the available data.

For details on how individual items should be maintained, the reader is referred to the chapter discussing the specific type of desalination plant; in addition, Chapter 7 discusses the avoidance of corrosion under nonoperating conditions.

9.1 Ambient Seawater

This section is dedicated to the equipment in contact with ambient temperature seawater, which does not exceed 40°C, and has been treated physically (with screens and filters) and chemically (disinfected). For more discussion on typical corrosion risks and related factors in ambient temperature seawater, refer to Chapters 4–6. In the following tables, a life estimation is given for each item with related corrosion risks and specific failure risks.

The tables given in the rest of the chapter use the following abbreviations for the different kinds of corrosion risks:

- C: Cavitation
- CC: Crevice corrosion
- CF: Corrosion fatigue
- CUC: Corrosion under coating
- DZ: Dezincification
- EC: Erosion-corrosion
- GC: General corrosion
- PC: Pitting corrosion
- SCC: Stress corrosion cracking

9.1.1 Evaporators

Table 9.1 shows the common materials used in evaporators and the main corrosion risk(s), along with the typical life.

TABLE 9.1 Evaporator materials and corrosion risks in aerated seawater.

PROCESS OR SECTION	ITEM	ALLOY	CORROSION RISK	LIFE (y)
MSF[a] Heat Rejection Section	Shell	C-steel	GC	30[b]
		Coated C-steel	CUC	10-40[c]
		Cu-Ni[d]	GC	40

a. MSF = Multistage flash
b. With suitable corrosion allowance
c. Coating defects significantly affect the lifetime
d. Cu-Ni = Copper-nickel

9.1.2 Heat Exchangers

Usually, heat exchanger failures appear in the form of tube leaks, where leaking tubes are plugged, and unit operation continues. Heat exchanger rehabilitation is decided when the unit water production drops significantly, which corresponds to plugging close to 30–40% of the tube bundle. Table 9.2 shows the common materials used in heat exchangers and the main corrosion risk(s), along with the typical life.

TABLE 9.2 Heat exchanger materials and corrosion risks in aerated seawater.

PROCESS OR SECTION	ITEM	ALLOY	CORROSION RISK	LIFE (y)
MSF Heat Rejection Section	Tubes	Ti Gr2[a]	None	Design life
		Cu-Ni	EC and PC	10-20[c]
	Tube sheets	Naval brass	DZ	10-20[d]
		NAB[b]	PC	10-20[e]
	Water box	Cu-Ni	PC	10-20
		Neoprene lined steel	CUC	10-15[f]
MSF Dump Condenser Section	Tubes	Ti Gr2	None	Design life
		66/30/2/2 Cu-Ni-Fe-Mn[h]	EC and PC	10-20[c]
		90/10 Cu-Ni	EEC and PC	10-20[c]
	Tube sheets	Naval brass	DZ	10-20[d]
		NAB	PC	10-25[e]
	Water box	90/10 Cu-Ni	PC	10-20
		Neoprene lined steel	CUC	10-15[f]

TABLE 9.2 (cont.) Heat exchanger materials and corrosion risks in aerated seawater.

PROCESS OR SECTION	ITEM	ALLOY	CORROSION RISK	LIFE (y)
MED[i] Final Condenser Section	Tubes	Ti Gr2	None	Design life
		Cu-Ni	EC and PC	10-20[c]
	Tube sheets	2205	PC and CC	10-20[g]
		316L	PC and CC	10-20[g]
	Water box	Stainless steel	PC	Design life[g]
		Neoprene lined steel	CUC	10-15[f]

a. Ti Gr2 = Titanium Grade 2
b. NAB = Nickel aluminum bronze
c. Depending on seawater cleanliness and velocity
d. Installation of sacrificial anodes should increase the lifetime
e. Sacrificial anodes and clean seawater should increase the lifetime
f. Coating defects significantly impact the lifetime
g. If sacrificial anodes fitted and changed as required
h. Cu-Ni-Fe-Mn = copper-nickel-iron-manganese
i. MED = Multiple effect distillation

9.1.3 Valves

Most desalination plant valves are the butterfly type. Different materials may be used in different parts of the plant. For more details see Chapters 4–6. The lifetime of valves is usually affected by three factors, i.e., seal integrity, environment, and duty factor (i.e., operation mode). Valves that are mostly closed can suffer more corrosion than one that is normally open. Table 9.3 shows some typical materials for the different valve components and the main corrosion risk(s). Also note that valve seals may require replacement during the valve life.

TABLE 9.3 Valve materials and corrosion risks in aerated seawater.

PROCESS OR SECTION	ITEM	MATERIALS	CORROSION RISK	LIFE (y)
MSF Make-up or Cooling Water Section	Body	Neoprene lined Steel	CUC	10-15[b, d]
		NAB	PC	10-20[c]
		Superduplex SS	None	Design life
	Disk	NAB	PC	10-20[c]
		Superduplex SS	None	Design life
	Shaft	Alloy 400	PC	10-20[c]
		Alloy K-500	PC	10-20[c]
		Superduplex	None	Design life
SWRO[a] Low Pressure Section	Body	Neoprene lined steel	CUC	10-15[b, d]
	Disk	NAB	PC	10-20[c]
	Shaft	Alloy K-500	PC	10-20[c]

TABLE 9.3 (cont.) Valve materials and corrosion risks in aerated seawater.

PROCESS OR SECTION	ITEM	MATERIALS	CORROSION RISK	LIFE (y)
SWRO High Pressure Section	Body	Superduplex SS	None	Design life
	Disk			
	Shaft			

a. SWRO = Seawater reverse osmosis
b. Coating defects significantly affect the lifetime
c. Provided seawater is free of pollutants
d. Life limited by life of rubber coating

9.1.4 Pumps

Table 9.4 shows some typical pump materials for the different components and the main corrosion risk(s), along with typical lifetimes. It is important to realize that not all pump alloys can be used in all combinations, i.e., a specific alloy cannot be picked for the pump case and then just any alloy is chosen for other components. For example, when superduplex is chosen, it is generally for all the pump components, while pumps with austenitic cast iron (ACI) cases generally have lower-alloy stainless steel internals. Chapter 4, Section 4.3 has more details on this. Also note that all pumps require replacement of seals, wear rings, and bearings at intervals throughout their lives.

TABLE 9.4 Pump materials and corrosion risks in aerated seawater.

PROCESS OR SECTION	ITEM	ALLOY	CORROSION RISK	LIFE (y)
SWRO Low Pressure Section	Case	Superduplex	None	Design life
		ACI	SCC	2-6[a]
		NAB	PC	15-25[b]
	Impeller	Superduplex	None	Design life
		CF8M	C	20[c]
		2205	C	20[c]
		NAB	PC	15-25[b]
	Shaft	Superduplex	None	Design life
		316L	None + ACI case	20[c]
		2205	None + ACI case	20[c]
		K-500	None + NAB case	15-25[d]
	Column Pipes	Superduplex	None	Design life
		ACI	SCC	2-6[a]
		NAB	PC	15-25[b]

TABLE 9.4 (cont.) Pump materials and corrosion risks in aerated seawater.

PROCESS OR SECTION	ITEM	ALLOY	CORROSION RISK	LIFE (y)
SWRO High Pressure Section	Case	6%Mo	None	Design life
		Superduplex	None	
	Impeller	6%Mo	None	
		Superduplex	None	
	Shaft	6%Mo	None	
		Superduplex	None	
MSF Makeup Section	Case	ACI	SCC	2-6[a]
		Superduplex SS	None	Design life
	Impeller	CF8M	C	Design life
		Duplex 2205	C	
		Superduplex SS	None	
	Shaft	316L	CF	Design life
		Duplex 2205	CF	
		Superduplex SS	None	
	Column pipe	ACI	SCC	2-6[a]
		316L	CC	20[e]
		Duplex 2205	CC	20[e]
		Superduplex SS	None	Design life
MED Supply Section	Case	Duplex SS	None if PREN > 40	Design life
	Impeller	Duplex SS	None if PREN > 40	
	Shaft	Duplex SS	None if PREN > 40	

a. Requires stress relief for longest life
b. If used in seawater free of pollutants and has postweld heat treatment (PWHT)
c. May be longer but it requires galvanic protection from ACI case
d. When used with an NAB pump only
e. Will suffer crevice corrosion at flanged joints unless cathodically protected

9.2 Brine

Brine in thermal desalination plants is characterized by a very low oxygen content (≤ 20 ppb), the presence of process chemicals (antiscalant, antifoaming, etc.), and a temperature that ranges from 40–120°C. The absence of oxygen and the closed system reduce the general corrosion rate of active alloys such as carbon steel.

9.2.1 Evaporators

Figure 9.1 shows thickness loss data obtained by ultrasound testing (UT) on an MSF flash chamber floor after 28 years of operation. For an original design thickness of 21.5 mm, there was a maximum thickness loss of 3% in the heat recovery and 5% in the heat rejection section. This would correspond to corrosion rates of 0.02 mm/y and 0.03 mm/y, respectively. Another contributing factor to the low thickness loss is the scale layer that confers additional protection to the carbon steel. Corrosion allowance reported in O&M manuals for carbon steel below the demister was overestimated in many designs and reached 12 mm in one case.

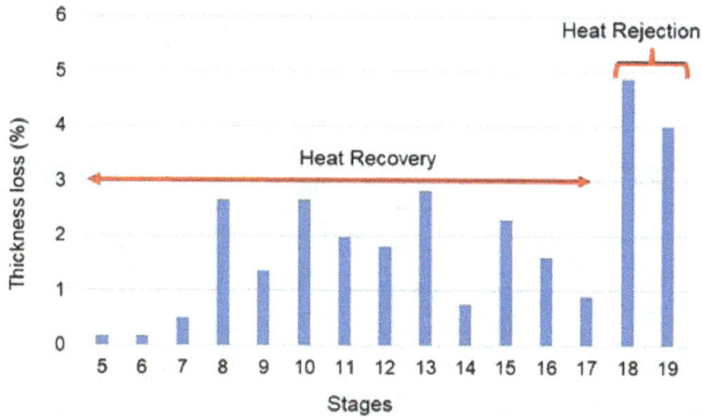

FIGURE 9.1 Thickness loss of MSF flash chamber floor after 28 years of operation.

In SWRO plants, the reject brine is fully aerated, at ambient temperature, and is at high pressure, so a high-alloy stainless steel with high strength is required (see Chapter 6). Table 9.5 shows the typical materials for the evaporator shell in thermal desalination plants and the main corrosion risk(s), along with the typical life.

TABLE 9.5 Evaporator materials and the main corrosion risks in deaerated brine.

PROCESS OR SECTION	ITEM	ALLOY	CORROSION RISK	LIFE (y)
MSF Heat Recovery Section	Shell	C-steel	GC	35[a]
		316L SS (solid or clad)	PC	Design life
MED All Effects Section	Shell	316L or lean duplex SS	PC	25
		Duplex SS 2205	PC	Design life

a. With suitable corrosion allowance

9.2.2 Heat Exchangers

Usually, heat exchanger failures appear in the form of tube leaks, where tubes are plugged, and unit operation continues. Heat exchanger rehabilitation is usually required when unit water production drops significantly, which corresponds to plugging close to 30–40% of the tube bundle. Table 9.6 shows the typical heat exchanger materials for the different components and the main corrosion risk(s), along with the typical life.

TABLE 9.6 Heat exchanger materials and the main corrosion risks in deaerated brine.

PROCESS OR SECTION	ITEM	ALLOY	CORROSION RISK	LIFE (y)
MSF Heat Recovery UC Section	Tubes	Ti Gr2	None	Design life
		90/10 Cu-Ni	EC	10-20[a]
	Tube sheets	Naval brass	DZ	10-20[b]
		NAB	PC	10-20[c]
	Water box	90/10 Cu-Ni	EC	10-20[a]
		Neoprene lined steel	CUC	10-15[d]
MSF Brine Heater Section	Tubes	90/10 Cu-Ni	PC	10-20[e]
		66/30/2/2 Cu-Ni-Fe-Mn	PC	15-25
	Tube sheets	Naval brass	PC/DZ	15-25
		NAB	PC	15-25
	Water box	90/10 Cu-Ni	PC	15-25
		Neoprene lined steel	CUC	10-15[d]
MED All Effects Section	Tubes	Ti Gr2	None	Design life
		90/10 Cu-Ni	EC	10-20
		Al brass	EC	10-20
	Tube Sheets	Lean duplex SS	None	15-25

a. Depends on clean seawater and correct water velocity
b. Requires sacrificial anodes to reduce corrosion
c. Requires clean seawater and sacrificial anodes with Ti tubes
d. Coating defects significantly affect the life
e. Depends on temperature and flow rate

9.2.3 Valves

Most desalination plant valves are the butterfly type. Different materials may be used in different parts of the plant. For more details see Chapters 4–6. The lifetime of valves is usually affected by three factors: seal integrity, environment, and duty factor (operation mode). Valves that are normally closed can be more prone to corrosion than those that are normally open. Table 9.7 shows some typical valve materials for the different components and the main corrosion risk(s), along with the typical life. Also note that valve seals may require replacement during the valve life.

TABLE 9.7 Valve materials and the main corrosion risks in deaerated brine.

PROCESS	ITEM	ALLOY	CORROSION RISK	LIFE (y)
MSF and MED Plants	Body	Cast Iron	GC	5-15[a, b]
		NAB	PC	10-20[a]
		CF8M	CC	10-20[b]
	Disk	NAB	PC	10-20[a]
		CF8M	CC	10-20[b]
	Shaft	Alloy 400	CC	15-25
		Alloy K-500	CC	15-25
		410 or 420 SS	CC	10-20[b]
		316L	CC	10-20[b]
SWRO	Body	Superduplex	None	Design life
	Disk			
	Shaft			

a. Provided pH is not too low

b. Provided dissolved oxygen is low

9.2.4 Pumps

Table 9.8 shows some typical pump materials used in brines, the main corrosion risk(s), and the typical life. It is important to realize that not all these alloys can be used in all combinations, i.e., a specific alloy cannot be picked for the case and then use just any alloy for the shaft and impeller. For example, when superduplex is chosen, it is generally for all the pump components, while pumps with ACI cases generally have lower-alloy stainless steel internals (see Chapter 4, Section 4.3 for details). Also note that all pumps require replacement of seals, wear rings, and bearings at intervals throughout their lives.

TABLE 9.8 Pump materials and the main corrosion risks in deaerated brine.

PROCESS OR SECTION	ITEM	ALLOY	CORROSION RISK	LIFE (y)
MSF and MED Brine Recycle and Brine Blowdown Sections (Low oxygen)	Case	Duplex 2205	None	Design life[a]
		ACI	SCC	10-20[b]
		NAB	PC	15-20[c]
	Impeller	Duplex 2205	None	Design life[a]
		CF8M	None	20[a, d]
	Shaft	Duplex 2205	None	Design life[a]
		316L	None	20[a, d]

TABLE 9.8 (cont.) Pump materials and the main corrosion risks in deaerated brine.

PROCESS OR SECTION	ITEM	ALLOY	CORROSION RISK	LIFE (y)
SWRO Brine Blow-down ERD Sections (fully aerated)	Case	Superduplex	None	Design life
	Impeller			
	Shaft			

a. Provided dissolved oxygen is low
b. Must be stress relieved
c. Must have PWHT
d. May be longer, but requires galvanic protection from ACI case.

9.3 Vapor Side

The main component affected by the vapor side corrosion, especially in MSF plants, is the vent piping (Chapter 4), although the evaporators may also experience problems.

9.3.1 Vent Piping

Table 9.9 shows some typical vent pipe materials, the corrosion risk, and a typical life. Where the temperatures are very high it may be necessary to upgrade to a 6%Mo austenitic or superduplex stainless steel. These alloys have a long life in the venting system.

TABLE 9.9 Vent piping materials and the main corrosion risks.

ITEM	ALLOY	CORROSION RISK	LIFE (y)
Venting Pipe	316L	SCC	5-25[a]
	2205	SCC	15-30

a. Depends on regular inspection and maintenance, and freedom from external SCC.

9.3.2 Evaporators

The MSF evaporator roof is subject to vapors that can be more-or-less corrosive to an extent; the corrosion rate can be determined by thickness measurement during periodic inspection. Figure 9.2 shows thickness loss data of carbon steel from an MSF evaporator after 28 years of service. Because of the low conductivity and very low oxygen concentration, the thickness loss did not exceed 4% of the original thickness, which was 15 mm. This value would correspond to a corrosion rate of 0.02 mm/y in the heat rejection section. The highest thickness loss is observed in the heat rejection section when oxygen is present in the flashed seawater. It should be noted that MSF evaporators were overdesigned at the time of construction (1980s), with a high corrosion allowance of 7 mm above the demister; the thickness can be optimized to reduce the evaporator cost.

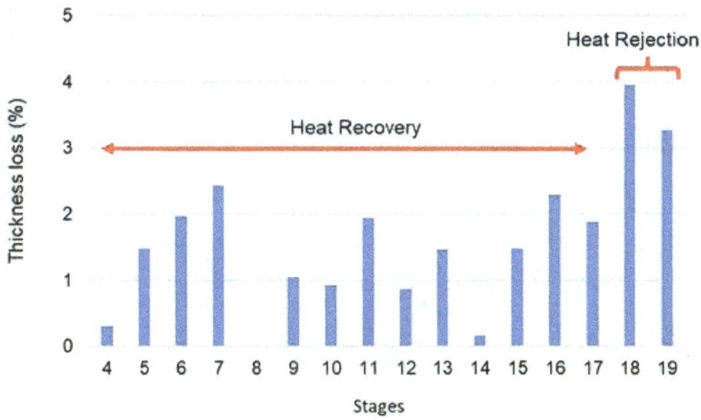

FIGURE 9.2 Thickness loss of MSF flash chamber roof after 28 years of operation.

References

1. Ghiazza, E., R. Borsani, and F. Alt, "Innovation in Multistage Flash Evaporator Design for Reduced Energy Consumption and Low Installation Cost," IDA Congress (Topsfield, MA: International Desalination Association, 2013).
2. Nada, N., and T. Hodgkiess, "Factors Affecting MSF Evaporator Life Expectancy," *Journal of Desalination and Water Reuse* 4 (2012): pp. 30–33.

10 Corrosion Assessment

In this chapter, different corrosion monitoring practices to assess the extent of corrosion under desalination plant operating conditions are briefly discussed. The monitoring can be direct, indirect, online, or offline. Depending on the corrosion form, suitable techniques, which may be qualitative or quantitative, are offered in the inspection equipment market. While monitoring and inspection are interrelated, some authors differentiate between them in terms of occurrence frequency. In all cases, inspection reveals corrosion damage and its extent, and when found, the item may require monitoring or rehabilitation (repair or replacement) to avoid any potential failure. The different corrosion risks are discussed in detail in previous chapters on the different desalination plant types, i.e., Chapters 4–6. Certain corrosion forms are difficult to monitor in a quantitative way and once they are detected, they should be corrected quickly. This is mainly the case for localized corrosion forms known for their unpredictability because of their rather random initiation points and propagation rates, for example SCC.

As for any industry, it is important to consider the cost of inspection as a significant part of the total cost due to corrosion for desalination plants. For each inspection or assessment method, there is a brief discussion about the best practices and limitations of the technique. The reader is advised to consult further references for details and limitations on the various corrosion monitoring techniques.[1–4]

Condition assessment of desalination plants is crucial from maintenance strategy and asset management perspectives. The recent introduction of artificial intelligence (AI) tools represents an opportunity that merits being explored for an optimization of the corrosion cost.[5] The focus of this chapter will be on the major components that significantly affect the reliability of the plant, as well as the maintenance cost.

10.1 Direct Assessment

A direct assessment of corrosion damage is necessary to monitor the evolving material degradation and to anticipate any potential failure. "Direct assessment" means that the corrosion damage has initiated and may be propagating on the metallic alloy. Table 10.1 summarizes the corrosion forms and their conventional inspection techniques together with related industrial standards, when applicable. While electrical resistance (ER) and linear polarization resistance (LPR) probes are proposed for monitoring, they are not suitable for assessing evaporator metal loss due to the specific conditions in them that hinder or challenge their corrosion monitoring ability.

TABLE 10.1 Typical corrosion forms observed in desalination and their corresponding monitoring techniques and recommended standards.

CORROSION FORM	INSPECTION TECHNIQUE OR METHOD	RECOMMENDED STANDARD
General Corrosion	Visual examination	EN 13018
	Thickness measurement by UT	ISO 16809
	Electrical Resistance (ER)	NA
	Linear polarization resistance (LPR)	NA
	Weight loss	ASTM G1
Pitting Corrosion	Visual examination	EN 13018
	Pit depth	ASTM G46
	Magnetic flux leakage (MFL)	ASTM E570
	Pulse eddy current	ASTM E2884
	Acoustic emission (AE)	EN 15856
	Field signature method	NA
Crevice Corrosion	Visual examination	EN 13018
	Pit depth	ASTM G46
SCC	Visual examination	EN 13018
	Dye penetrant test (DPT)	ISO 3452
	Phase array UT (PAUT)	NA

NA = Not Applicable

The monitoring of general corrosion is usually performed offline by thickness measurement using the ultrasonic test method (UT), because it does not require high skills to analyze the data. Thickness measurements are mainly practiced in multistage flash (MSF) desalination plants. Inspectors focus on the evaporator where the measurements of the thickness are conducted every 2–3 years at different locations. The locations are selected to be representative of brine, splash, and vapor areas. Figure 10.1 shows an example of thickness measurement location in an MSF flash chamber floor and roof. Weight loss coupons and LPR methods are also used but mainly for research purposes. The other materials (stainless steel and copper alloys) have a good general corrosion resistance, which excludes the need to monitor corrosion for these materials. On the other hand, sacrificial anodes are also subject to an assessment for their general corrosion in terms of their weight and dimensions during the planned shutdown. Typically, sacrificial anodes are replaced when they have lost more than 50% of their original thickness.

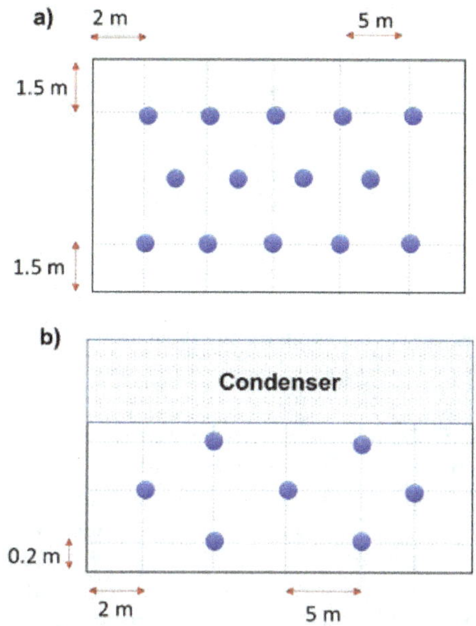

FIGURE 10.1 Thickness measurement test points: (a) MSF flash chamber floor, and (b) MSF evaporator roof (internal).

Pitting corrosion assessment is usually limited to its detection offline on limited and critical equipment. This equipment includes heat exchangers in thermal desalination plants, which are inspected with limited frequency because of the related cost and large number of heat exchanger tubes. Eddy current (EC) is the main technique adopted where pitting of cupronickel tubes can be detected and quantified. Figure 10.2 shows a typical example of eddy current mapping of the thickness loss, and it also highlights any pitting present. In addition to EC testing for assessing pitting or crevice corrosion, these can also be visually assessed using a pit depth gauge. This is done during preventative maintenance performed during planned shutdowns, where pitting is suspected or has previously been observed, because tubes must be removed for this type of inspection. This is also the case for pumps, valves, and pipes made of corrosion resistant alloys (CRAs). Testing for crack detection using dye penetrant testing (DPT) or phased array ultrasonic testing (PAUT)-may also be carried out during planned shutdowns. Such cracks could develop into corrosion fatigue or stress corrosion cracking (SCC) in pump shafts, venting pipes, or welds.

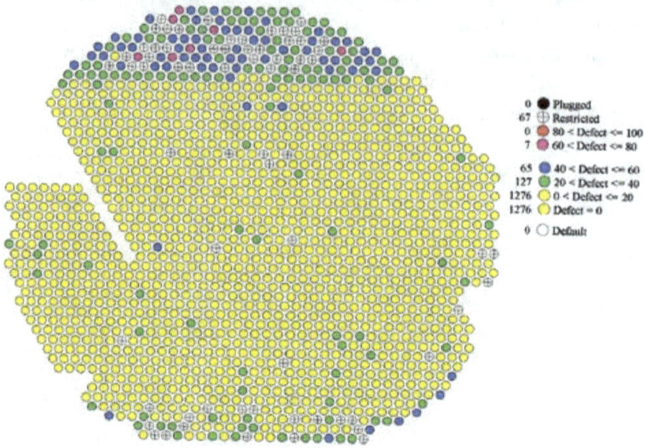

FIGURE 10.2 Eddy current testing of MSF stage 1 (heat recovery) after 21 years of operation.

The limitations of corrosion monitoring techniques are well discussed in the literature by many authors.[3, 4] For instance, ER and LPR probes are designed based on uniform corrosion, which makes them unsuitable for localized corrosion detection. Then, any change of pH or oxygen that may initiate localized corrosion attack cannot be detected, unless their location is chosen with care. Moreover, practical experience has shown discrepancies between ER data and weight loss corrosion data, and fluctuation of LPR data in field conditions.[6] Taking the example of an MSF evaporator or water box, ER or LPR probes will suffer from scale deposition, which will hinder the probe response and produce unreliable corrosion data. ER probes can be installed in selected vapor area locations to detect any oxygen ingress that can initiate a corrosion process. This can be deduced from the usefulness of ER for similar applications to monitor atmospheric corrosion.[7, 8] For localized corrosion, many efforts have been undertaken to combine techniques such as LPR with electrochemical techniques such as electrochemical noise (EN) or harmonic distortion analysis (HDA).[9] However, these combinations remain at the level of laboratory research and have not reached a maturity to be validated for long-term field conditions. In the case of SCC, two approaches can be considered: (1) monitoring of environmental conditions leading to SCC, or (2) installing a prestressed probe loaded to the 0.2% proof stress. The efficacy of such a probe will be determined by its location. The most desirable locations are not always available for the installation of probes.

Monitoring coating degradation can be considered mostly as qualitative, although coating damage is categorized and well defined in industrial standards. Table 10.2 summarizes some selected coating damage types as categorized in ISO 4628[10] and their inspection methods with recommended standards. The relevant standards give further details on the inspection requirements and data interpretation.

TABLE 10.2 Typical coating damage observed in desalination plants, their corresponding assessment techniques, and recommended standards.

DAMAGE FORM	INSPECTION METHOD	RECOMMENDED STANDARD
Blistering	Visual comparison with pictorial standards	ISO 4628-2 or ASTM D714
Rusting		ISO 4628-3 or ASTM D610
Cracking		ISO 4628-4
Flaking		ISO 4628-5
Delamination Degree	Measurement of delamination width around a defect	ISO 4628-8

10.2 Indirect Assessment

Environmental monitoring–in terms of critical factors that can affect corrosion–is one of the widely implemented approaches in desalination plants. Dissolved oxygen, pH, temperature, velocity, etc., are strictly monitored to avoid initiating severe corrosion or fouling (inorganic scale, biofouling) processes. The parameters are assessed at well-defined process locations to check both corrosion risk and equipment performance such as for chlorination, deaeration, and acidification. The fluids that are chiefly monitored include feedwater, brine, and product water (distillate in thermal desalination, and permeate in SWRO). Table 10.3 summarizes the monitored environmental parameters, related risk, and frequency of monitoring. When abnormal values are noticed, further laboratory analysis can be conducted, such as the analysis for copper or iron concentration to investigate a potential corrosion source in the process. In some thermal desalination plants, a weekly monitoring of metallic cations (iron [Fe] and copper [Cu]) concentration is adopted for brine and product water (distillate). High iron contents imply corrosion of the evaporators, while high copper contents imply corrosion of the heat exchanger tubes.

Other than environmental monitoring, rotating equipment is also subject to vibrational monitoring to anticipate the risk of corrosion fatigue or SCC. This is very critical for pumps handling various fluids in desalination plants, where monitoring is performed using a variety of methods. If the operator detects abnormal sounds, a vibration meter should be used to check for deleterious vibration that can lead to catastrophic failure. Vibrational analysis requires specific skills to identify the vibration source and fix it properly.

TABLE 10.3 Environmental monitoring parameters practiced in desalination plants.

DESALINATION PROCESS	CORROSION PARAMETER	CORROSION RISK*	MONITORING FREQUENCY
All	Residual chlorine	MIC, PC, CC	Once every 8 hours or continuous
	Velocity	EC, PC, MIC	Continuous
	Electrical conductivity	GC, PC, CC	Continuous
Thermal	Temperature	PC, SCC	Continuous
MSF	Dissolved oxygen	GC, PC, CC, BC	Once every 8 hours or continuous
SWRO	pH	PC, CC	Once every 8 hours or continuous
	Redox	CC	Continuous
	Coagulant ($FeCl_3$)	CC	Continuous

MIC = microbially influenced corrosion, PC = pitting corrosion, CC = crevice corrosion, EC = erosion corrosion, SCC = stress corrosion cracking, GC = General corrosion, BC = Bimetallic (galvanic) corrosion; $FeCl_3$ = ferric chloride

References

1. Roberge, P.R., *Corrosion Inspection and Monitoring* (New York, NY: Wiley Interscience, 2007).
2. Yang, L., *Techniques for Corrosion Monitoring*, 2nd ed. (Cambridge, United Kingdom: Woodhead Publishing, 2020).
3. NACE International Task Group T-3T-3, "Techniques for Monitoring Corrosion and Related Parameters in Field Applications" (Houston, TX: NACE International, 2020).
4. Tan, Y., K. Lepkova, S. Bailey, and S.R. Gubner, "A Critical Review on Various Corrosion Testing and Monitoring Techniques," in *Proceedings of the Australasian Corrosion Association Conference: Corrosion and Prevention*, D. Nicholas, ed. (Coffs Harbour, Australia: Australian Corrosion Association, 2009): p. 044.
5. Scully, J.R., and P.V. Balachandran, "Future Frontiers in Corrosion Science and Engineering: Part III: The Next Leap Ahead in Corrosion Control May be Enabled by Data Analytics and Artificial Intelligence," *Corrosion* 75 (2019): pp. 1395–1397.
6. Jarragh, A., S. Al-Sulaiman, A. Al-Shamari, B. Lenka, M. Islam, and S. Parkash, "Evaluation of the Effectiveness of Online Corrosion Monitoring Utilizing ER/LPR Probes and Coupon within Hydrocarbon Systems," Corrosion 2014, paper no. 3934 (Houston, TX: NACE International, 2014).
7. Xia, D., S. Song, Z. Qin, W. Hu, and Y. Behnamian, "Review–Electrochemical Probes and Sensors Designed for Time-Dependent Atmospheric Corrosion Monitoring: Fundamentals, Progress, and Challenges, *Journal of the Electrochemical Society* 167, 3 (2020): p. 037513.

8. Wade, S.A., K.J. Begbie, and A. Trueman, "Measuring Atmospheric Corrosion of Industrial Infrastructure Using Electrical Resistance Corrosion Sensors," Corrosion Control Conference (Victoria, Australia: Australian Corrosion Association, 2007).
9. Heselmans, J., K. Hladky, M. Holdefer, and R. Wessels, "New Corrosion Monitoring Probe Combines ER, LPR, HDA, Floating B-Constant, Electrochemical Noise and Conductivity Measurements," Corrosion 2013, paper no. 2332 (Houston, TX: NACE International, 2013).
10. ISO 4628 (latest edition), "Paints and Varnishes–Evaluation of Degradation of Coatings–Designation of Quantity and Size of Defects, and of Intensity of Uniform Changes in Appearance" (Geneva, Switzerland: International Standards Organization).

11 Standards, Quality, and Procurement

When material is being procured, either for new build or repairs, it is important to recognize it is often necessary to ensure its quality to properly fulfill its intended function. Within this framework, quality assurance (QA) and quality control (QC) practices are recommended with clear responsibility assignment in the supply chain. It is common to request the contractor to provide a QC plan with relevant standards.

In this chapter, some gaps in certain material standards are identified and related guidance is given as a support on how to define further materials specifications to ensure they will be fit for purpose. The advisability of additional tests with some alloys is also discussed. The importance of QA and QC (QA/QC)–both with vendors and on delivery–is emphasized, as are the important factors affecting satisfactory fabrication. Although such technical specifications and QA/QC add cost, it is far less costly than that of a major failure.

11.1 Standards and Specifications

11.1.1 Metals

When procuring metals and alloys it is common to use American Society for Testing and Materials (ASTM) specifications, such as ASTM A240 for stainless steel plate and sheet. However, these are product standards written largely by material producers to achieve minimum acceptable requirements with special focus on the chemical composition, mechanical properties, and a brief mention about suitable heat treatment. No details are provided about how to achieve the best properties, or how to test to ensure the product has been properly manufactured. With the worldwide increase in the number of corrosion resistant alloy (CRA) vendors, and increasing project desire to minimize costs, it is essential that additional testing, over and above that required by ASTM, is written into a supplementary specification to ensure the supply of good quality material.

11.1.2 Polymers

Advances in polymer chemistry, and the understanding of how to blend and formulate polymers, allow for very specific grades of a particular polymeric material to be produced. It is almost never sufficient to specify polymeric material generically, and the grade of polymer must be matched to the end-use application. For example, polypropylene is commonly used for handling caustic solutions, but the block copolymer (grade H2222) offers better chemical degradation and cracking resistance than the homopolymer. In general, for many chemical applications, unfilled polymers without plasticizers, have the optimum chemical resistance. If other grades are used, then exposure testing of the intended material in the target environment will provide confirmation that the material is acceptable. Many ASTM standards cover the plastic (or its polymeric derivatives) specification, testing of its properties, and its safe utilization. This is extremely important when considering that most polymeric materials will be in contact with aggressive and hazardous process chemicals.

11.2 Importance of QA/QC

11.2.1 Metals and Alloys

Be aware that some alloys have limits on the available thickness. Duplex stainless steel is a good example. It is not currently possible to quench duplex stainless steel thicker than 250 mm to avoid formation of sigma phase near the center.[1] Heat treatment of duplex stainless steel needs careful control; witnessing the testing is advisable, unless the supplier has been fully audited. It is also important that QA/QC monitoring is maintained throughout the manufacture and supply. This costs money but the cost remains lower than a major failure. Customers want to lower costs, and producers with limited experience may cut corners without realizing the implications.

Many companies rely heavily on two standards produced by Standards Norway (NORSOK): M-630 for materials, and M-650 for vendor approvals.[2,3] NORSOK M-650 was developed as a way of prequalifying manufacturers of "difficult" materials (largely duplex and superduplex stainless steels) to ensure that previous costly delays in production of components in these alloys were avoided. ISO[(1)] 17782 is a successor to NORSOK M-650, and this is also a vendor approval document.[4] However, just because a vendor claims to be NORSOK or ISO 17782 approved, be wary and check the validity of the certificate. NORSOK certification lasts five years but auditing should be done every three or four years, because manufacturers often change things in the production cycle or install new equipment that has not been audited. Auditing can be expensive, but it usually pays for itself when items are needed in a hurry, such as for urgent repairs.

1. ISO = International Standards Organization

Also check what the company is approved to supply. This includes both product forms and thickness range. There is usually a good reason if they do not have approval over the full thickness range. It is essential to ensure that any potential supplier is approved for all needed material requirements.

Another document that is not as widely used as it should be is EEMUA 218,(2) which gives details on how to obtain good quality duplex stainless steels.[5] Without going into details, it has a number of main points:

- NORSOK M-650 preferred (now also ISO17782), but good historic data is acceptable
- Audits recommended more frequently than every five years
- UKAS(3) accredited test laboratories only (now also covered by ISO 17025)
- All testing at the center, at one-half thickness (1/2T)
- All parts of the manufacturing route and heat treatment must be covered in detail
- Surveillance and witnessing during manufacture
- A technical compliance audit required, particularly for new suppliers, over and above a QA audit
- Metal stockists to be treated in the same way as manufacturers, with the same responsibilities (material bought from inventory to be subject to independent 3rd party testing to verify the mill certificate properties)
- Positive material identification (PMI) and Feritscope(4) checks required

In addition to procurement specifications and monitoring during manufacture, it is worthwhile checking material on delivery with some simple nondestructive testing. Hardness tests can sometimes be useful, and Francis et al. describe the use of the Feritscope to assess the quality of duplex stainless steels.[6] They comment on the factors that affect the readings and show that significant amounts of sigma or chi phase produce a significant reduction in the ferrite reading. It is possible to take a lot of readings very quickly in many locations; looking at the range of the readings suggests when there could be precipitates.

- Readings >40 are OK
- Readings from 30–39 could show problems if inconsistent, and further investigation is required
- Readings <30 are a strong indication of sigma phase

Suspect material should then be checked with nondestructive or destructive metallography, as there are causes other than sigma that can give low ferrite readings, such as poor pickling.[6]

In addition to this, there are simple nondestructive tests that can be applied on delivery to try and identify problems early on. One of these is the use of random PMI checks, using a portable X-ray analyzer, to check the composition of all CRAs.

---

2. EEMUA = Engineering Equipment and Materials User Association
3. United Kingdom Accreditation Service
4. Feritscope is a registered trademark of Fischer Instrumentation, Leipzig, Germany.

Although these instruments are expensive (but the costs are coming down), you only have to identify one batch of rogue material to have saved the cost of the analyzer. One of the authors (Francis) has been involved in some failure analyses where the supposedly high alloy stainless steel supplied was, in fact, 316L. PMI quickly identifies errors of this kind on delivery. Alloy mix up is a significant issue in the mills, stockholders, distributors, and fabricators. Color coding at the mill is a good way of avoiding alloy mix ups.

With nickel alloys, and particularly heat treatable, aged alloys, a simple hardness test can confirm that the alloy is in the correct heat treatment condition. If it is not in the correct condition, this could lead to the further aging of aged material, which could significantly reduce ductility, or it could lead to the non-aging of solution annealed material, which would have too low mechanical strength.

The use of a simple gauge to measure relative magnetic permeability of 90/10 copper-nickel (Cu-Ni) is discussed in Section 11.3.2.2 below, and this can be used as a quick check on delivery.

With many of these requirements it is often simpler to have QA/QC staff, or a third party witness, at the suppliers to verify that everything has been done correctly to the technical specification. Although this adds to costs, it is far less than that of a major failure, such as that described in Reference 7, where the cost was estimated to be millions of dollars.

11.2.2 Polymers

Polymer formulations may differ significantly between manufacturers, so for demanding applications polymers are often specified by manufacturer's trade name and specific grade. Acceptance tests can include Fourier transformation infrared spectroscopy (FTIR) or differential scanning calorimetry (DSC) done on samples of batches of material. Specific gravity tests can also be useful to differentiate between polymers. More information on the typical standards adopted to determine the quality of polymers and their related products is given in Table 11.1.

TABLE 11.1 Typical standards followed for QC of polymers and their products.

PARAMETER	STANDARD	MATRIX	METHOD
Epoxy Content	ASTM D1652	Epoxy resins	Titration
Total Acid Number	ASTM D1386 EN ISO 2114	Polyester resins, paints	Titration
Hydroxyl Value	EN ISO 2554	Unsaturated polyester resins	Titration
Water	ASTM 6869	All polymers	KFT[a]
Thermostability	ISO 182-3	Vinyl chloride homopolymers and copolymers	Oxidation stability

a. Karl-Fischer titration

For fiberglass reinforced plastic (FRP) in contact with seawater, the use of epoxy or vinyl ester resins is usually adequate, but it is important to specify an ultraviolet (UV)-resistant external coating and additional measures if the pipe is aboveground. Whatever resin is specified, because FRP pipe is mostly exposed to strong sunlight in desalination plants, testing for outdoor exposure is necessary, as described by ASTM D4364.[8] Moreover, resin and glass quality tests are often defined in the project specifications with related standards.

11.3 Test Requirements

This section contains some suggestions for testing required to ensure the supply of good quality metallic alloys, based on the authors' and published experiences.

11.3.1 Austenitic Cast Iron (ACI)

There are a number of grades of ACI, which are classified into two families: flake and spheroidal graphite alloys. Because of their better seawater corrosion resistance and ductility, spheroidal grades are widely used in desalination plants. Typical grades include alloys D2 or D2B (ASTM A439), offering good seawater corrosion resistance, while retaining some weldability. As discussed in Chapter 4, all ACI alloys will suffer from stress corrosion cracking (SCC) in warm seawater because the stresses at the interfaces between the different particles in the structure are sufficient to initiate cracking. The initiation of SCC is delayed by ensuring that all castings are fully stress relieved. The stress relief can be done by heating to 600–650°C at a rate of 50–100°C/hr.[9] Thickening the castings also reduces stresses and delays failure, but with the high nickel content, this introduces a cost penalty.

11.3.2 Copper Alloys

11.3.2.1 Brasses

In desalination plants, the principal brass alloy is aluminum brass; it is used for heat exchanger tubes in older MED plants, and in the lower temperature heat recovery section in multistage flash (MSF) plants. In addition to the usual tests for dimensional tolerances, straightness, ammonia cracking, etc., it is important to test for residual carbon films from the manufacturing process. These films are well known to cause pitting of copper tubes in certain fresh waters, and many authors have also seen failures of other copper alloy heat exchanger tubes in seawater due to carbon films.[10, 11]

Many standards, such as ASTM B111[12] for seamless copper alloy heat exchanger tubes, do not make any mention of carbon films or other deleterious films in the bore. The current European standard for copper alloy heat exchanger tubes, EN

12451, does not specifically require that heat exchanger tubes should be free from films in the bore, but leaves it up to the purchaser.[13] It is important to specify that samples of tube will be tested using a thermal test, such as in the LECO[(5)]-type test. In this combustion test, the sample is heated in a stream of oxygen and the CO_2 produced at different temperatures is determined. By heating at increasing temperatures, the resulting carbon originating from various sources can be isolated as follows:

- 200°C for solvents
- 650°C for types of grease
- 850–900°C for carbon films

Only the carbon measured at the highest temperature is used for assessing the tube quality. The pass/fail criterion that has been adopted is the same as that used for copper tubes for use in fresh waters, i.e., carbon ≤ 0.2 mg/dm^2. Tubes that fail this test can be cleaned by internal grit blasting or by pickling, and then retested.

Also, naval brass with its high strength has been used for tube sheets with copper alloy tubes, and although it may dezincify, the tube sheet thickness (≥ 35 mm) and the slow propagation rate in deaerated conditions makes leakage unlikely.

Manganese bronze is actually a high tensile brass, and this will dezincify rapidly in seawater. If a high strength copper alloy is required, then it should be nickel aluminum bronze (NAB), although nickel alloy K-500 and duplex stainless steel are also used where high strength is required. The additional tests for these alloys are discussed below.

11.3.2.2 Copper-Nickel Alloys

Copper-nickel alloys being used for heat exchanger tubes should be subject to the same tests as aluminum brass, as discussed in the previous section, including the LECO test for carbon films, with the same pass/fail criteria.

The 90/10 copper-nickel (Cu-Ni) being used as plate, sheet, or pipe may also suffer from another problem. The alloy gives the best corrosion resistance, especially from the impingement attack point of view, when the iron content is in the range 1.5–2.0 wt%.[14] However, all the common standards for the alloy give the iron limit as 1.0–2.0 wt% (ASTM) or 1.2–2.0 wt% (ISO). To ensure adequate resistance to erosion-corrosion, it is necessary to specify that the minimum iron content must be 1.5 wt%.

This level of iron exceeds the solubility limit for the alloy at room temperature. However, if the iron is not in solution, the erosion-corrosion resistance decreases quite markedly.[14] If the alloy is not cooled fast enough from the annealing temperature, iron may precipitate. Determination of the magnetic permeability with

5. LECO is a registered trademark of the LECO Corporation, St. Joseph, Michigan, USA

a simple device, such as the Severn Gage[(6)] Low Mu Permeability Indicator model is recommended, at the same intervals as the other tests (e.g., for carbon films and residual stresses). Material with a magnetic permeability of >1.10 should be rejected.

11.3.2.3 Nickel Aluminum Bronze (NAB)

NAB is a widely used alloy in both cast and wrought forms. NAB has sometimes been observed to suffer dealuminification, more commonly on castings and particularly under crevice corrosion conditions.[15] It has been found that keeping the nickel content ≥ 0.5% above the iron content will prevent dealloying.[16] This is not covered by most specifications and should be added as a requirement when specifying this alloy, in either cast or wrought form.

At the high temperatures used in hot working or during welding, NAB has a duplex structure that is alpha and beta. During cooling the beta phase transforms to alpha and kappa. However, if the cooling rate is too fast, the beta phase does not fully transform.[16] Beta phase is highly anodic to the alpha phase, and it is selectively and rapidly attacked in seawater.

It is recommended when ordering NAB that it be given a heat treatment after the final hot operation (welding, rolling, etc.). For castings and welds, it is recommended that the Royal Navy heat treatment of 675° ± 25°C for 2–6 hr followed by air cooling is specified. For wrought products (such as thick plate) a heat treatment of 725°±25°C for the same period followed by air cooling is recommended.[10]

11.3.3 Stainless Steels

11.3.3.1 Austenitic Stainless Steels

It is sometimes worth requiring an ASTM A262 Method E[17] test for 300 series alloys to check their susceptibility to intergranular corrosion attack (IGA). IGA on austenitic stainless steels is known to occur when carbide precipitation (sensitization) is expected because of a heating process (operating conditions or welding operation). This phenomenon is directly related to the carbon content and can be avoided by specifying L grade for a carbon content ≤0.03%. The alternative is to do a full analysis by, for example, optical emission spectroscopy, as most PMI machines cannot detect carbon.

The higher alloy austenitic stainless steels, such as the 6%Mo grades, should have the additional tests required by NORSOK standard M-630.[2] These are a microsection (suitably etched and viewed at X500), to prove that the structure is free of sigma phase, and an ASTM G48 Method A corrosion test at 50°C for 24 hr.[18] The pass criteria are no pitting and a weight loss less than 4 g/m^2. This ensures that

6. Severn Gage is a registered trademark of the Severn Engineering Co., Inc., Coppell, Texas, USA

there are no deleterious phases or chromium-depleted zones, due to poor pickling. It is important to test samples with the same finish that will go into service.

11.3.3.2 Duplex Stainless Steels

There have been some well publicized problems with duplex and superduplex stainless steels due to poor manufacturing procedures.[7, 19-21] ASTM A923 addresses the detection of sigma phase in duplex stainless steels, but it has been criticized for not rejecting material that might fail in service.[22] The oil and gas industry has produced ISO 17781 as a set of (metallurgical, mechanical, and corrosion) tests to guarantee the quality of duplex stainless steels of all types, and in different product forms as castings, wrought product, and hot isostatic pressing (HIP).[23] It is recommended that these tests be adopted for the procurement of duplex stainless steels by all industries. There are three tests, the first of which is a microsection to detect the presence of third phases and measure the phase balance. The second is a Charpy impact toughness test with different test temperatures and pass/fail criteria for the different duplex grades. The third is a corrosion test to ASTM G48[18] or A1084[24] (for lean duplex) with different test temperatures for the different grades. These tests will detect not just sigma phase, but also other deleterious phases, such as nitrides.

The impact toughness test guarantees that, even after welding, there will be adequate toughness for most applications. Because third phases reduce toughness, it is also a measure of the quality of the microstructure. Passing the corrosion test shows that deleterious third phases are not present in sufficient amounts to significantly affect the corrosion resistance or are not present at a sufficiently large diameter to sustain a stable pit. The corrosion test can also detect nitrides, which may cause pitting or just a high weight loss.[25]

The microsection enables the phase balance to be determined and to see any third phases, if the sample is polished and etched correctly.[26] This is particularly helpful if the material has failed test 1 or 2, because it helps diagnose the nature of the problem, identify possible root causes, and establish remedial action. If a duplex alloy contains too many third phase precipitates, or precipitates above a certain size, it would be impossible to pass either the impact toughness test or the corrosion test. The efficacy of these tests is demonstrated by some case histories in Reference 26.

11.3.4 Nickel Alloys

The specifications for many older nickel alloys permit relatively high carbon levels, because they were written before the widespread introduction of argon oxygen decarburization (AOD) refining. Similarly to stainless steels, nickel alloys with higher carbon contents may have grain boundary precipitates that can lower corrosion resistance and lead to intergranular corrosion. Carrying out an ASTM G28 method

A test will determine whether the alloy has been suitably manufactured and heat treated[27]. The pass-fail criteria need to be determined by comparing the corrosion rate for properly annealed material, as described in the references in ASTM G28[27].

Table 11.2 shows the corrosion rate just sufficient for intergranular corrosion in the G28A test for some common nickel alloys.[28] However, from a commercial point of view this is just the threshold, and a common fail in this test is a corrosion rate >1mm/y, except for C-276, which is often >12mm/y. Note that these rates are for wrought material. Welds and cast products can have a higher corrosion rate in a satisfactory condition and pass/fail criteria must be established for these separately.

TABLE 11.2 Corrosion rates of some common nickel alloys sufficient to just show intergranular corrosion in the ASTM G28 method A test.[19]

ALLOY	UNS NUMBER	TEST DURATION (hr)	CORROSION RATE (mm/y)
20	N08020	120	0.61
825	N08825	120	0.46
625	N06625	120	0.46
C-276	N10276	24	9.14

The alternative is to do a full analysis by, for example, optical emission spectroscopy, as most positive material identification (PMI) machines cannot detect carbon.

Alloy K-500 may be supplied in the annealed or the aged condition. If components are to be machined, then it is easier in the annealed condition and aging must be carried out subsequently to obtain the required strength. Such requirements need to be included in the technical specification.

11.3.5 Titanium Alloys

The properties of titanium alloys are very susceptible to a reduction caused by small changes in the content of oxygen, hydrogen and, particularly, iron. Be prepared to check these independently if you are not familiar with the supplier. High residuals can result in an alloy that is unweldable or has reduced corrosion resistance. Where SCC risk or corrosion fatigue (cyclic stresses) are expected oxygen, nitrogen and carbon should be limited to the extra low interstitial (ELI) content in the grade to give better fatigue resistance, and it is important to check that this has been specified and delivered. Thermal processing of titanium alloys at temperatures exceeding ~500°C can result in the formation of alpha case, which is a brittle surface layer of diffused-in oxygen. If alpha case exists in a layer greater than 3–5 μm thick it should be removed, and it should be totally removed for applications involving cyclic stresses, to improve fatigue life.

11.4 Castings

The above discussion has principally addressed wrought products, but poor foundry practice can lead to solidification cracks, cold cracks, high levels of segregation, shrinkage, porosity, local high stresses or an undesirable microstructure, all of which may lead to premature failure. When selecting a foundry for CRA castings, particularly larger ones, there are some factors to take into account:

- What is their experience? Many CRA's require some skill to cast correctly; it is always best to use a foundry with prior experience.
- Do they have computer modeling for solidification and stress distribution? The design of running, gating and positioning of feeders or padding can affect the porosity, residual stress and tear formation in castings. The ability to model this before pouring any metal can save a lot of money and time.[29]
- Is the heat treatment furnace adequate? The importance of this is discussed in Reference 30.
- Is the quench tank satisfactory (adequate sample to volume ratio, cooling and agitation)? See References 29 and 30.
- Can they do fast transfers (<60 seconds)? See References 29 and 30.

Poor heat treatment can also cause problems with duplex stainless steel castings.[29, 30] The additional tests in ISO 17781 also cover castings. When procuring duplex castings it is usual to start with ASTM A995 and add the extra test requirements from ISO 17781 under the testing options[31].

The authors have looked at numerous failures of duplex castings caused by poor quality due to a lack of adequate specifications and a lack of suitable QA/QC.

Castings in NAB should also have a heat treatment as the final stage after any weld repair and before machining, to optimize both the strength and the corrosion resistance. The recommended heat treatment is 675° ±25°C for 2–6 hours (depending on thickness), followed by air cooling.[16]

With castings in all alloys, especially large ones, it is recommended that a map be required showing all the sites where weld repair has been carried out and what postweld heat treatment (PWHT) was carried out, preferably with a third party witness. The importance of PWHT for NAB was discussed above, and ASTM A995 requires solution annealing after major weld repair on all duplex stainless steel castings.[31]

11.5 Fabrication

Having purchased the correct alloys, they will usually require fabrication. Ideally a fabricator with a proven track record in welding the alloy(s) of concern should be used, as some CRA's require a level of skill that needs time to learn, e.g., titanium.

Fabrication in the shop is preferred to site welding for CRAs, because it is much easier to control the welding conditions and cleanliness. Where sufficiently skilled site welders are in short supply, it has proved effective with some seawater reverse osmosis (SWRO) plants to spool the piping in a shop, and ship to site, where most of the assembly is by mechanical jointing.

When welding many CRAs, there will inevitably be some heat tint in the heat-affected zone (HAZ). With stainless steels and nickel alloys, the thin layer of oxide has taken chromium from the matrix, which now has a reduced corrosion resistance under the heat tint. The heat tint should be removed either by abrasion, such as a stainless steel wire brush, or a suitable pickling paste. Mechanical grinding should be avoided, because it can introduce microcrevices into the metal surface, which are preferential sites for the initiation of crevice corrosion.[32] If iron or steel debris is left as contamination after cleaning, it can give rise to corrosion of lower alloy stainless steels (300 series), and it can produce unsightly staining on high alloyed grades, after coming in contact with moisture (Figure 11.1).

FIGURE 11.1 **Rusting due to iron contamination from poor cleaning of superduplex stainless steel welds.** ***Photo courtesy of Rolled Alloys***

It is important to remove the heat tint on the side that is exposed to the process fluid where possible, particularly for 300 series stainless steels. These grades are the most susceptible to corrosion under the heat tint because of their lower alloy content. If piping is being spooled, it can be feasible to get the spools pickled at a cleaning plant with large pickling baths. The authors have seen this done on some SWRO projects.

Chapter 2 discusses the welding of each alloy type, but for some CRAs there are tests that should be carried out in addition to those specified in ASME IX, or whatever fabrication standard is being followed. These should be carried out on both the procedure qualification record (PQR) and weld repair PQR for castings and the individual welder qualifications. These were discussed in Chapter 2 and are summarized in Table 11.3. For more detail see Chapter 2, Section 2.2.

TABLE 11.3 Additional tests for CRA welds.

ALLOY	FILLER	TESTS IN ADDITION TO DESIGN CODE
90/10 Cu-Ni	90/10 Cu-Ni	Magnetic permeability
	70/30 Cu-Ni	None
NAB	Cu/9Al/3Fe	Test for adequate strength
	NAB	Micro for residual beta
300 Series Stainless Steel	Matching	None
	309	Check delta ferrite
6%Mo Austenitic SS	625	Micro to check for sigma ASTM G48 at 50°C
Duplex SS	Overalloyed (higher Ni)	ISO 17781
Ni-Cr-Mo	Ni-Cr-Mo	Micro to check for sigma ASTM G48 at 50°C
Titanium	Ti	Check color Micro for alpha case

11.5.1 Nickel Alloys

NORSOK standard M-601 recommends the same tests for Ni-Cr-Mo alloys, such as alloy 625, as specified for 6%Mo austenitic stainless steels.[33] That is a suitably etched microsection at X500, to show the absence of deleterious phases, and an ASTM G48 test at 50°C for 24 hr. There should be no pitting and a weight loss <4 g/m^2. The samples should have the same surface finish as is to be used in service.

11.5.2 Titanium Alloys

Titanium alloys should ideally be free from weld discolouration. Welders must ensure not only that the actual weld pool is shielded but that the hot weld is maintained in an inert atmosphere until it has cooled to well below 400°C. Oxidation above 450°C results in a transformation to alpha case at the surface and as a result surface ductility, fatigue strength and corrosion resistance are reduced. If such a layer is found it must be removed. The reader is directed to an excellent summary of the issue and a color comparator by Talkington et al.[34]

11.5.3 Polymers

Joints in FRP piping systems are generally glued for tapered joints up to 150 mm diameter pipe, and butt and strap joints are used for larger diameters. Figure 11.2

shows some typical tapered joints awaiting assembly. In some applications butt and strap joints may be used for smaller diameter pipes. It is critical that the workers who are employed to make these joints are trained and experienced. Typically, the pipe manufacturer will supply joint kits, which consist of precut veil and reinforcement that can be applied at site by suitably trained and competent personnel. In remote locations, it is common for the pipe vendor to send a supervisor to site. The supervisor trains the local workforce to perform the joining work.

FIGURE 11.2 FRP pipe with tapers for jointing. ***Photo courtesy of Steve Clarke***

With FRP pipe, flanged joints are required to connect to other equipment, such as valves, and the flanges are usually attached to the pipes by butt and strap connections. Flange to pipe connections are typically made in the pipe shop and final, on-site joints are in the pipe runs.

References

1. Francis, R., *The Corrosion of Duplex Stainless Steels: A Practical Guide for Engineers* (Houston, TX: NACE International, 2018).
2. NORSOK M-630 (latest edition), "Material Data Sheets for Piping," Edition 5 (Oslo, Norway: Standards Norway, 2010).
3. NORSOK M-650 (latest edition) "Qualification of Manufacturers of Special Materials," Rev. 4 (Oslo, Norway: Standards Norway, 2011).
4. ISO 17782 (latest edition), "Petroleum, Petrochemical and Natural Gas Industries–Qualification of Manufacturers of Special Materials" (Geneva, Switzerland: International Standards Organization, 2018).
5. EEMUA Publication 218, "Quality Requirements for the Manufacture and Supply of Duplex Stainless Steels" (London, United Kingdom: Engineering Equipment and Materials Users Association, 2010).

6. Francis, R., G. Warburton, G. Byrne, and J. Wilson, "QA/QC Tools to Ensure the Quality of Duplex Fittings and Other Components," Duplex Stainless Steel 2010 (Zutphen, Holland: KCI, 2010).
7. Rorvik, G., L.A. Marken, I.M. Kulbotte, M. Aursand, S. Olsen, and K.S. Karlsen, "Influence of Intermetallic Precipitates on the Mechanical Properties and Environmental Cracking Resistance of Duplex Stainless Steel Fittings–A Case History (Part 1)," Duplex Stainless Steels 2010 (Zutphen, Holland: KCI, 2010).
8. ASTM D4364 (latest edition), "Standard Practice for Performing Outdoor Accelerated Weathering Tests of Plastics Using Concentrated Sunlight" (West Conshohocken, PA: ASTM International, 2013).
9. Nickel Institute, "Properties and Applications of Ni-resist and Ductile Ni-resist Alloys," Technical Paper 1231 (Toronto, Canada: Nickel Institute, 1975).
10. Francis, R., *The Corrosion of Copper and Its Alloys: A Practical Guide for Engineers* (Houston, TX: NACE International, 2010).
11. Giuliani, L., and R. Cigna, "Corrosion Problems of Copper Alloys in Desalination Plants," *Desalination* 22, (1977): pp. 379–384.
12. ASTM B111 (latest edition), "Standard Specification for Copper and Copper-Alloy Seamless Condenser Tubes and Ferrule Stock" (West Conshohocken, PA: ASTM International).
13. EN 12451:2012, "Copper and Copper Alloys. Seamless, Round Tubes for Heat Exchangers" (Pilsen, Czech Republic: European Standards, 2012).
14. Bailey, G.L., "Copper-Nickel-Iron Alloys Resistant to Seawater Corrosion," *Journal of the Institute of Metals* 79 (1951): pp. 243–292.
15. Richardson, I., "Guide to Nickel Aluminium Bronze for Engineers," C. Powell, ed., Publication No. 222 (Hemel Hempstead, United Kingdom: Copper Development Association, 2016).
16. Campbell, H.S., "Aluminium Bronze Alloys Corrosion Resistance Guide," Publication No. 80, July 1981 (Hemel Hempstead, United Kingdom: Copper Development Association, 1981).
17. ASTM A262 (latest edition), "Standard Practices for Detecting Susceptibility to Intergranular Attack in Austenitic Stainless Steels" (West Conshohocken, PA: ASTM International).
18. ASTM G48 (latest edition), "Standard Test Methods for Pitting and Crevice Corrosion Resistance of Stainless Steels and Related Alloys by Use of Ferric Chloride Solution" (West Conshohocken, PA: ASTM International).
19. Mahajanam, S., H. Rincon, and D. McIntyre, "Metallurgical Examination of Defects in Duplex Stainless Steel Pipe Fittings," *Materials Performance* 49, 4 (2010): p. 56.
20. Ryengen, E., and C. Wintermark, "Lessons Learned from Heat Treatment of Components in 22Cr and 25Cr Duplex Stainless Steel (and Other Materials)," Duplex Stainless Steels 2010 (Zutphen, Holland: KCI, 2010).
21. Turbeville, E., F. Busschaert, S. Benum, and B. Hadsen, "A Summary of Recent Experiences on Designing and Fabricating Subsea Oil and Gas Production Systems with 22Cr and 25Cr Piping Materials," Duplex Stainless Steels 2010 (Zutphen, Holland: KCI, 2010).

22. ASTM A923 (latest edition), "Standard Test Methods for Detecting Detrimental Intermetallics in Duplex Austenitic/Ferritic Stainless Steels" (West Conshohocken, PA: ASTM International).
23. ISO 17781 (latest edition), "Petroleum, Petrochemical and Natural Gas Industries—Test Methods for Quality Control of Microstructure of Ferritic/Austenitic (Duplex) Stainless Steels" (Geneva, Switzerland: International Standards Organization, 2017).
24. ASTM A1084 (latest edition), "Standard Test Method for Detecting Detrimental Phases in Lean Duplex Austenitic/Ferritic Stainless Steels" (West Conshohocken, PA: ASTM International).
25. Byrne, G., R. Francis, and G. Warburton, "Variations in Mechanical Properties and Corrosion Resistance of Different Alloys Within the Generic Designation UNS S32760," Duplex America 2000 (Zutphen, Holland: KCI, 2000).
26. Byrne, G., "Meaningful Tests for the Quality of Duplex Stainless Steels," paper no. 10876, Corrosion 2018 (Houston, TX: NACE International, 2018).
27. ASTM G28 (latest edition), "Standard Test Methods for Detecting Susceptibility to Intergranular Corrosion in Wrought Nickel-Rich, Chromium Bearing Alloys" (West Conshohocken, PA: ASTM International).
28. Baboian, R., *Corrosion Tests and Standards: Application and Interpretation*, ASTM Manual Series MNL 20, 2nd ed. (West Conshohocken, PA: ASTM International, 2005).
29. Francis, R., and S. Hebdon, "The Selection of Stainless Steels for Seawater Pumps," paper no. 5446, Corrosion 2015 (Houston, TX: NACE International, 2015).
30. Francis R., and S. Hebdon, "The Limiting Thickness for Duplex Stainless Steels," paper no. 3651, Corrosion 2014 (Houston, TX: NACE International, 2014).
31. ASTM A995 (latest edition), "Standard Specification for Castings, Austenitic-Ferritic (Duplex) Stainless Steel for Pressure-Containing Parts" (West Conshohocken, PA: ASTM International).
32. Tossey, B.M., A. Agrawal, and J. Shingledecker, "Effect of Post Weld Clean-up Method on the Localized Corrosion of Duplex Stainless Steels," paper no. 4384, Corrosion 2014 (Houston, TX: NACE International, 2014).
33. NORSOK M-601, Rev. 06 (latest edition), "Welding and Inspection of Piping" (Oslo, Norway: Standards Norway, 2015).
34. Talkington, J., D. Harwig, H. Castner, and G. Mitchel, "Development of Titanium Weld Colour Inspection Tools," *Welding Journal* (March 2000): pp. 35–38.

Nominal Composition of Some Alloys Commonly Used In Desalination Plants

Table A.1 Austentic Cast Irons—all of the alloys below contain a small magnesium addition (typically ~0.06%) to spheroidize the graphite.

ASTM A439 Designation	UNS NUMBER	NOMINAL COMPOSITION (wt%)						
		Fe	C*	Ni	Cr	Si	Mn	P*
D-2	F43000	Bal	3	20	2	2.3	1	0.08
D-2B	F43001	Bal	3	20	3.3	2.3	1	0.08
D-2C	F43002	Bal	2.9	22.5	0.5*	2.5	2.1	0.08
D-2M	F43010	Bal	2.7	22.5	0.2*	2	4.2	0.08
D-3	F43003	Bal	2.6	30	3	2.2	1*	0.08
D-3A	F43004	Bal	2.6	30	1.2	2.2	1*	0.08
D-4	F43005	Bal	2.6	30	5	5.5	1*	0.08
D-5	F43006	Bal	2.4	35	0.1*	2.1	1*	0.08
D-5B	F43007	Bal	2.4	35	2.5	2.1	1*	0.08

Bal = Balance
*Maximum

A.2 Copper Alloys

TYPE	COMMON NAME	UNS NUMBER	NOMINAL COMPOSITION (wt%)								
			Cu	Zn	Ni	Fe	Mn	Al	Sn	Pb	Other
WROUGHT	Al-Brass	C68700	77	Bal	-	-	-	2	-	-	As
	Muntz Metal	C28000	60	Bal	-	-	-	-	-	0.2	-
	Naval Brass	C46400	61	Bal	-	-	-	-	1	-	-
	DZR Brass	C35330	62	Bal	-	-	-	-	-	2	As
	Mn-Bronze	C67500	58	Bal	-	0.8	0.4	-	0.7	0.1	-
	90/10 Cu-Ni	C70600	Bal	-	10	1.7	1	-	-	-	-
	70/30 Cu-Ni	C71500	Bal	-	30	0.7	0.7	-	-	-	-
	66/30/2/2 Cu-Ni-Fe-Mn	C71640	Bal	-	30	2	2	-	-	-	-
	Nibron	C72400	Bal	-	14.5	1.5	0.3	2	-	-	Mg
	Al-Bronze D	C61400	Bal	-	-	2.5	-	7	-	-	-
	Al-Bronze E	C63000	Bal	-	5	3	-	10	-	-	-
	Nickel Aluminum Bronze	C63200	Bal	-	4.5	4	-	9	-	-	-
CAST	Admiralty Gunmetal	C92400	Bal	2	-	-	-	-	10	2	-
	Phosphor Bronze	C90700	Bal	-	-	-	-	-	10	-	P
	Leaded Bronze	C93700	Bal	-	-	-	-	-	10	10	P
	LG2 Gunmetal	C83600	Bal	5	-	-	-	-	5	5	-
	LG4 Gunmetal	C92410	Bal	2	-	-	-	-	7	3	-
	M-Bronze	C92200	Bal	4	-	-	-	-	6	1.5	-
	Al-Bronze	C95200	Bal	-	-	3	-	8.5	-	-	-
	Nickel Aluminum Bronze	C95800	Bal	-	4.5	4	-	9	-	-	-

Bal = Balance

TABLE A.3 The nominal compositions of some commonly used wrought stainless steels.

TYPE	UNS NUMBER	GENERIC NAME	NOMINAL COMPOSITION (wt%)									PREN#
			Fe	C*	Cr	Ni	Mo	N	Cu	W	Mn	
Austenitic	S30403	304L	Bal	0.03	18	8	-	-	-	-	1.5	18
	S31603	316L	Bal	0.03	17	10	2	-	-	-	1.5	24
	S31703	317L	Bal	0.03	19	12	3	-	-	-	1.5	29
	N08020	Alloy 20	Bal	0.07	20	33	2.5	-	3	-	1.5	28
	N08904	904L	Bal	0.02	20	25	4	-	1.5	-	1	34
Super Austenitic	S31254	6%Mo	Bal	0.02	20	18	6	0.2	0.7	-	0.7	43
	N08367	6%Mo	Bal	0.03	20	25	6	0.2	-	-	1	43
	N08926	6%Mo	Bal	0.02	20	25	6	0.2	1	-	1	43
Ferritic	S41003	410	Bal	0.03	12	0.7	-	-	-	-	1	12
Martensitic	S42000	420	Bal	0.2	13	-	-	-	-	-	0.7	13
Super Ferritic	S44735	Al 29-4C	Bal	0.03	29	0.3	4	-	-	-	0.5	42
	S44660	Seacure	Bal	0.03	26	2.5	3	-	-	-	1	36
Lean Duplex	S32304	2304	Bal	0.03	23	4	0.3	0.10	0.3	-	1.5	24
	S32101	2101	Bal	0.03	21	1.5	0.3	0.22	-	-	4.5	26
	S82011	2102	Bal	0.03	21	1.5	0.3	0.22	-	-	2.5	26
	S32202	2202	Bal	0.03	22	2	0.3	0.2	-	-	1.5	26
Duplex	S32205	2205	Ba	0.03	22	5	3.2	0.17	-	-	1	35
	S31803	2205	Bal	0.03	22	5	3	0.13	-	-	1	34
Superduplex	S32760	Z100	Bal	0.03	25	7	3.5	0.25	0.7	0.7	0.7	>41
	S32750	2507	Bal	0.03	25	7	3.6	0.27	-	-	0.8	>41

PREN =%Cr + 3.3(%Mo + 0.5 ×%W) + 16 × %N

Bal = Balance

* = Maximum

TABLE A.4 The nominal compositions of some commonly used cast stainless steels.

TYPE	ASTM[§]	GENERIC NAME	NOMINAL COMPOSITION (wt%)								PREN[#]
			Fe	Cr	Ni	Mo	N	Cu	W	Mn	
Austenitic	CF8M	316	Bal	17	10	2	-	-	-	1	24
	CK3MCuN	6%Mo	Bal	20	18	6	0.2	0.7	-	0.8	43
	CN3Mn	6%Mo	Bal	21	25	6	0.2	0.3	-	1	43
Standard Duplex	Grade 4A	2205	Bal	22	5.5	3	0.16	0.5	-	1	35
25%Cr Duplex	Grade 1A[¶]	CD4MCu	Bal	25	5	2	-	3	-	0.7	32
	Grade 1B	CD4MCuN	Bal	25	5	2	0.14	3	-	0.7	34
Superduplex	Grade 5A	2507	Bal	25	7.5	3.7	0.25	-	-	1	>41
	Grade 6A	Z100	Bal	25	8	3.5	0.25	0.7	0.7	0.7	>41

§ = ASTM designations taken from A351 for austenitic and A995 for duplex

PREN = %Cr + 3.3(%Mo + 0.5×%W) + 16×%N

Bal = Balance

¶ = Withdrawn

A.5 Nickel Alloys

NAME	UNS NUMBER	NOMINAL COMPOSITION (wt%)									PREN*
		Fe	C	Ni	Cr	Mo	Cu	Al	Ti	W/Nb	
Alloy 400 Alloy K-500	N04400 N05500	2 1.5	0.05 0.05	65 65	- -	- -	Bal Bal	- 3	- 0.6	- -	N/A N/A
Alloy 825	N08825	Bal	0.03	40	21	3	2	-	0.9	-	31
Alloy 28	N08028	Bal	0.02	31	27	3	1	-	-	-	37
Alloy 625	N06625	3	0.05	Bal	21	9	-	-	-	3.5Nb	51
Alloy C-276	N10276	5	0.01	Bal	15.5	16	-	-	-	3W	72
Alloy C-22	N06022	3	0.01	Bal	22	13	-	-	-	3W	70
Alloy 59	N06059	1	0.01	Bal	23	15.5	-	0.3	-	-	74
Alloy 2000	N06200	1	0.01	Bal	23	16	1.5	0.25	-	-	76
Alloy 686	N06686	3	0.01	Bal	21	16	-	-	0.1	3.5W	79
Alloy 718	N07718	Bal	0.05	52	19	3	-	0.5	0.9	5Nb	29
Alloy 925	N09925	Bal	0.03	44	21	3	1.8	0.3	2.1	-	31
Alloy 725	N07725	Bal	0.02	56	20	8	-	0.2	1.5	3Nb	46

Bal = Balance
*PREN = %Cr + 3.3(%Mo +0.5×%W) + 16×%N
N/A = Not Applicable

A.6 Titanium Alloys

TYPE	GRADE	UNS NUMBER	NOMINAL COMPOSITION (wt%)							
			Ti	O_2*	Mo	Ni	Al	V	Pd	Ru
Alpha	1	R50250	Bal	0.18	-	-	-	-	-	-
	2	R50400	Bal	0.25	-	-	-	-	-	-
	3	R50550	Bal	0.35	-	-	-	-	-	-
	7	R52400	Bal	0.25	-	-	-	-	0.15	-
	11	R52250	Bal	0.18	-	-	-	-	0.15	-
	12	R53400	Bal	0.25	0.3	0.8	-	-	-	-
	16	R52402	Bal	0.25	-	-	-	-	0.05	-
	17	R52252	Bal	0.18	-	-	-	-	0.05	-
	26	R52404	Bal	0.25	-	-	-	-	-	0.1
Alpha-Beta	5	R56400	Bal	0.20	-	-	6	4	-	-

Bal = Balance
* Maximum

Index

Page numbers followed by *f* indicate figures, those followed by *t* indicate tables.

www.ingramcontent.com/pod-product-compliance
Lightning Source LLC
LaVergne TN
LVHW022226200926
848481LV00006B/2135
* 9 7 8 1 5 7 5 9 0 4 2 2 1 *